Kepler's Cosmological Synthesis

History of Science and Medicine Library

VOLUME 39

Medieval and Early Modern Science

VOLUME 20

The titles published in this series are listed at brill.com/hsml

Kepler's Cosmological Synthesis

Astrology, Mechanism and the Soul

By
Patrick J. Boner

BRILL

LEIDEN • BOSTON
2013

Cover illustration: Kepler's Supernova, SN 1604, appears as a new star in the foot of Ophiuchus near the letter N. In: Johannes Kepler, *De stella nova in pede Serpentarii*, Prague: Paul Sessius, 1606, pp. 76–77. Courtesy of the Department of Rare Books and Manuscripts, Milton S. Eisenhower Library, Johns Hopkins University.

Library of Congress Cataloging-in-Publication Data

Boner, Patrick, author.
Kepler's cosmological synthesis: astrology, mechanism and the soul / by Patrick J. Boner.
pages cm. — (History of science and medicine library, ISSN 1872-0684; volume 39; Medieval and early modern science; volume 20)
Based on the author's doctoral dissertation, University of Cambridge, 2007.
Includes bibliographical references and index.
ISBN 978-90-04-24608-9 (hardback: alk. paper) — ISBN 978-90-04-24609-6 (e-book)
1. Kepler, Johannes, 1571–1630—Philosophy. 2. Cosmology—History. 3. Astronomy—History. I. Title. II. Series: History of science and medicine library; v. 39. III. Series: History of science and medicine library. Medieval and early modern science; v. 20.

QB36.K4.B638 2013
523.1092—dc23

2013013707

This publication has been typeset in the multilingual "Brill" typeface. With over 5,100 characters covering Latin, IPA, Greek, and Cyrillic, this typeface is especially suitable for use in the humanities. For more information, please see www.brill.com/brill-typeface.

ISSN 1872-0684
ISBN 978-90-04-24608-9 (hardback)
ISBN 978-90-04-24609-6 (e-book)

This book is printed on acid-free paper.

To María Fernanda

CONTENTS

ACKNOWLEDGEMENTS

This book is based on my doctoral dissertation, which I completed at the University of Cambridge in 2007. I wish to thank my supervisor, Nicholas Jardine, for his service and support. A constant source of compassion and insight, Nick has contributed to this work in countless ways. Liba Taub, my advisor, led me to identify new links between Kepler and the classical tradition. I also wish to thank Miguel Ángel Granada (University of Barcelona) and Peter Barker (University of Oklahoma), who kindly hosted me at different points in my doctoral program. I am grateful to Miguel Ángel and Peter for their continuing role in my research.

My interest in the history of science was first awakened by Robert A. Hatch at the University of Florida. When Bob first agreed to supervise my undergraduate thesis on Kepler's astrology, he encouraged me to consider it a book from the very beginning. Bob has welcomed the burden of editing more than a decade of writing while witnessing it evolve into the form that it now takes in the following pages. No one familiar with Bob's work (and wit) will fail to recognize his indelible influence. I am grateful for his invaluable assistance and assume full responsibility for any remaining errors and omissions.

This book has been thoroughly revised over the course of two postdoctoral fellowships. I wish to thank the Humboldt Foundation for the chance to explore the wealth of resources at the Bavarian Academy of Sciences and the Bavarian State Library. My time in Munich was marked especially by the collegiality of Daniel A. Di Liscia, who encouraged me to examine the correspondence of Kepler more closely. I also wish to thank the National Science Foundation for the opportunity to come to the Johns Hopkins University to complete this book. My advisor, Lawrence M. Principe, offered sound advice and suggested ever new sources to better and broaden my scope of study. A vibrant community of scholars at the Department of History of Science and Technology provided critical comments that greatly improved this work. I am especially grateful to Sharon Kingsland and Danielle Stout for making me feel more at home by providing me with the perfect facilities.

The final stages of revision were made possible by funding from the Spanish Ministry of Science and Innovation, as part of the Project FFI2009-07156 "Cosmología, teología y antropología en la primera fase de la Revolución Científica (1543–1633).” I wish to thank my fellow project members, Miguel Ángel Granada and Dario Tessicini, for permission to reproduce a large portion of my paper, “Kepler's Vitalistic View of the Heavens: Some Preliminary Remarks” (2012), in Chapter 1. I also wish to thank Christoph Lüthy and two anonymous referees for revising an earlier version of the manuscript. As with any project that takes multiple years to complete, there are many others to thank. Most of you know who you are, and I am forever in your debt.

Finally, this book would not have been possible without the loving support of my family and friends. In Baltimore, the Hopkins Harriers provided a vital distraction and daily remedy from the doldrums of writing, not to mention a stronger sense of humility by outrunning the author regularly. I am also grateful to the group of colleagues and friends whose lunch gatherings are now modestly known as the “Meeting of the Minds.” Their comradery and keen sense of humor lifted my spirit and gave me greater focus. I dedicate this book to my wife, María Fernanda, who has supported me at every stage of the project. The completion of the original manuscript coincided happily with the birth of our beautiful daughter, Annabel.

LIST OF ILLUSTRATIONS

LIST OF ABBREVIATIONS

JKGW KEPLER, Johannes, *Gesammelte Werke*. Ed. Max Caspar and Walther von Dyck *et al.* 22 vols. Munich: C. H. Beck, 1937–.

TBOO BRAHE, Tycho, *Opera omnia*. Ed. J. L. E. Dreyer *et al.* 15 vols. Copenhagen: Swets & Zeitlinger, 1972.

INTRODUCTION

> First, know that heaven and earth and the liquid fields, the moon's bright sphere and Titan's star, a spirit within sustains; in every limb mind moves the mass and mingles with the mighty frame.[1]
>
> —Virgil, *Aeneid* 6: 724–727

The recent discovery of extra-solar planets has created a wave of revolutionary changes in the earth and planetary sciences. In the wake of these changes, the boundaries between 'earth scientist' and 'planetary scientist' have been blurred by a surge of new studies that suggest their intersection. Yet blurred boundaries are nothing new to science. Astronomers may point to Johannes Kepler (1571–1630) as a pioneer of this interplanetary vision. In his *Dream* (1634), an imaginary voyage to the moon, Kepler applied his knowledge of natural philosophy to describe what life might look like on the lunar surface. By projecting the laws of perspective, Kepler also explained why a lunarian observer might reasonably conclude that the moon stood motionless at the center of the world. The analogy was clear. "Everyone declares that the motion of the heavenly bodies around the earth [is] evident to the eyes," Kepler wrote, "but to the eyes of the lunarians, it is evident that our earth rotates while their moon is motionless."[2] To press the imagination further, Kepler suggested that some of the lunar beings who enjoyed this view roamed the moon nomadically, either on foot, by air, or "on boats, following the receding waters."[3]

Known as an early form of science fiction, Kepler's *Dream* prompts readers to marvel at the idea of alien beings crossing the lunar surface "in large hordes" and diving to "the far depths of the sea" to avoid the heat of the sun.[4] At the heart of this imaginative tale, however, is a serious attempt to understand the physical reality of a faraway body. Today, astronomers and marine biologists cross similar lines when they contemplate life beneath the surface of Europa and Jupiter's other icy satellites. This cognitive leap has been elaborated by Stephen Hawking, who claims

[1] Virgil portrays this spirit as similar in nature to fire and the universal source of life.
[2] *JKGW*, 11,2, 354.23–25. Cf. Rosen, 1967, p. 106.
[3] *JKGW*, 11,2, 330.19–20. Cf. Rosen, 1967, p. 27.
[4] *JKGW*, 11,2, 330.23–26. Cf. Rosen, 1967, p. 28.

that life on the ocean floor could hold clues for alien ecosystems similar to our own.[5] It has even been introduced to the general public by James Cameron in his documentary, "Aliens of the Deep," where we encounter a new world of life inhabiting the volcanic vents. While Kepler could never have imagined such a world, he plumbed the analogical depths of the earth to similar effect. Thus, the biblical account of creation in the earth's waters served to illuminate the origin of comets in the celestial ether. In the same way that "great whales" had been born in the sea, Kepler claimed, the ether gave rise to comets "in every corner" of the heavens.[6] Kepler would later turn to human anatomy to explain the complementary roles of the sun and the planets in the body of the cosmos. If the sun was "a sort of heart," the earth was "the liver or spleen."[7] Kepler blurred disciplinary boundaries because he believed the stars were subject to change in some of the same ways as the sublunar sphere. "Copernicus granted the right of citizenship in the heavens to the earth," he declared,[8] and with it came a new way of knowing the world of mutability beyond the moon.

How did Kepler account for change according to the heliocentric hypothesis? The answer reveals a forgotten feature of his world view and a critical part of the new cosmology. Kepler envisioned a world of change beyond the perennial motion of the planets. Comets and new stars cried out for explanation in his "far-reaching philosophy of the heavens."[9] The constant cycle of decay and renewal that he imagined beyond the earth echoed the ancient principle of perfection that involved continual change. According to Aristotle, God had "fulfilled the perfection of the universe by making coming-to-be [on earth] uninterrupted."[10] This constancy of creation and corruption secured a state of perpetual being that allowed the world thus to participate in perfection. Kepler saw no reason to limit this principle to any part of his world picture. The new planetary status of the earth suggested the same form of physical explanation extending to the fixed stars. While Kepler continued to study closely the light conveyed by

[5] In his television series, "Into the Universe," Hawking suggests that underwater aliens on Europa "would probably swim in a similar way to our own ocean life, since liquid water is the same stuff everywhere."

[6] *JKGW*, 4, 59.1–10. Cf. Genesis 1: 20–22.

[7] *JKGW*, 6, 416.31–32.

[8] *JKGW*, 1, 246.23–24.

[9] Westman, 2011, p. 317.

[10] Martin, 2011, p. 42.

the stars,[11] the earth now became the basis for physical reasoning. Change occurred everywhere and the power of analogy prevailed.

Kepler focused on the four causes of Aristotle in his physical reasoning. He assigned a number of natural faculties as efficient causes to explain how change happened in the heavens. As a student of Aristotle, Kepler knew that a natural faculty was often found in a living being. He variously borrowed and broke from Aristotle without ever fully rejecting his natural philosophy.[12] From plants and animals to the nutritive and perceptual powers of human beings, a faculty fulfilled a particular function in a particular part of the body. Nature had given four faculties "to account for nutrition," for example, "[namely] the attractive, retentive, digestive, and expulsive."[13] Kepler argued that the earth employed a similar set of faculties for gulping down sea water and digesting it deep below the surface. The earth acted as "the body of a sublunar soul" whose faculties served the same functions as many other living beings.[14] When Kepler extended this analogy and assigned the same faculties to the heavens, their basis in the living body was never lost entirely. Kepler claimed the same faculty that kept the fluid clear around the eye also acted in the celestial ether, where it "purged impure vapors and preserved the pellucidity of the ether."[15] Whether he accounted for a comet or another celestial novelty, the body provided a powerful metaphor for explaining cosmic change. It is an example of the fundamental place of analogy and metaphor in the broader enterprise of early modern inquiry.[16] By bridging the heavens and earth, Kepler redefined the role of astronomy in relation to natural philosophy and human knowledge.

Astronomers continued to sharpen this view over the seventeenth century, fitting the heliocentric hypothesis with a fully uniform account of change. For Johannes Hevelius (1611–1687), the ether was nothing more than a finer form of air. "It is rightly inferred," Hevelius wrote, "that the same essence prevails wherever the same subtle and diaphanous matter

---

[11] On the role of starlight as a singular resource in astronomy, see Muir-Harmony, DeVorkin, and Abrahams, 2012, pp. 195–197.

[12] Kepler echoes the earlier chorus of critical voices such as Copernicus, who "had seen Aristotelian principles rejected, transformed, and adapted in countless ways." See Goddu, 2010, p. 331.

[13] *JKGW*, 16, no. 494, 49–51.

[14] Schwaetzer, 1997, p. 236.

[15] *JKGW*, 1, 269.17–31.

[16] In his recent survey of the Scientific Revolution, Lawrence M. Principe laments "the loss of the comprehensive early modern vision." See Principe, 2011, pp. 134–135.

abounds."[17] Since the same material medium extended "from our earth to the farthest stars," Hevelius claimed that generation and corruption could occur "everywhere in the ether."[18] Hevelius drew deeply from Kepler, following his physical theory as diligently as he deployed his mathematics. He argued that the very matter surrounding the earth and the stars filled the full expanse of the universe. Such an essential continuity allowed for the complete unification of celestial and terrestrial physics. The sublunar sphere enjoyed "a great affinity" with the heavens and shared "many of the same inclinations and passions [*inclinationes et passiones*]."[19] Hevelius could not identify any major difference that amounted to more than a matter of measure. Even the surface of the sun, "the supreme celestial globe," was assigned an earthly analogy in the form of "boiling water or some very hot liquid."[20] At the core of this continuity, Hevelius identified "a natural faculty of generation and corruption" that acted as the author of new things it later destroyed.[21] Such a sequence of decay and renewal recalled the cyclical course of the weather. Hevelius believed that "a natural force [*vis naturalis*]" brought about "new bodies in the heavens" just as the earth generated and then dissipated "certain conditions of the weather by absorbing the matter that [made] them up."[22] Hevelius further explained that comets came from celestial exhalations that collected together in condensed areas of the ether. After they reached a point of perfection, comets soon began to break apart. "When a comet has moved beyond that mature age, as it were, and begun to grow old," Hevelius wrote, it dissolved into the same ethereal vapors that first gave place to it.[23] As the century came to a close, Isaac Newton (1642–1727) would spin similar speculations that shaped his cosmological legacy through the Enlightenment.

Kepler's comparison of the heavens and earth is often associated with his "mechanical explanation" of planetary motion.[24] By applying the principles of terrestrial mechanics to the paths of the planets, Kepler expressed an early form of the mechanistic world view "that has increasingly defined modern science."[25] According to this view, Kepler represents

---

17 Hevelius, 1668, p. 355.
18 Ibid.
19 Ibid.
20 Ibid., p. 359.
21 Ibid., p. 358.
22 Ibid.
23 Ibid., p. 383.
24 Hooykaas, 1987, p. 466.
25 Keller and Brummer, 2002, p. 264.

an early stage in seventeenth-century astronomy that found fruition in Newton, who fit "the Copernican model of the universe definitively into a mechanistic system of nature."[26] Such a conception has been described by some historians as "the death of nature" and the replacement of an organic model with a system that was far more pliable and predictable.[27] For others, the shift in cosmic metaphor reflects "the pervasive influence of machines" on early modern life and the role that such a view served to restore law and order to a society ravaged by wars of religion.[28] In modern ecology, it has even caught the attention of advocates of "a postmechanistic ethic" who aim to overturn "the manipulation of agro-ecosystems."[29] By re-assessing industrial agriculture in the light of a variety of non-economic values, these voices call for alternatives to the cause-and-effect relationships of mechanistic science. Kepler is among a small group of thinkers whom these authors credit with our current view of nature as "a grand and exquisite machine."[30] The flaw in this account is that Kepler claimed that comets and other forms of celestial change were entirely unpredictable. When it came to understanding an event such as a new star, Kepler counseled, "let every man [be] a liar."[31]

Advocates of the modern reductionist view confront an even greater irony when they portray Kepler as the promoter of purely mechanical principles. In their call for "the recognition of aesthetic and even spiritual [values]" in modern agriculture, they appeal to the very sort of view that Kepler entertained throughout his career.[32] Kepler argued that the earth was ensouled, and this informed his explanation of celestial mutability when he turned to the earth as a source of comparison. We gain a better sense of this association by examining Kepler's account of spontaneous generation in "the liquid fields" of the ether.[33] To explain how the heavens gave birth to new forms, Kepler compared their creative activity to the cycle of decay and renewal that occurred on earth. In the same way that a steady stream of plants and animals arose from putrid matter, the production of comets and new stars served to purge the heavens of

[26] Hooykaas, 1987, p. 463.
[27] Merchant, 1989.
[28] Oelschlaeger, 1991, p. 77.
[29] Keller and Brummer, 2002, p. 268.
[30] Ibid., p. 264.
[31] *JKGW*, 1, 292.1–3. Cf. Romans 3:4.
[32] Keller and Brummer, 2002, p. 270.
[33] *JKGW*, 1, 268.21. On the meaning of this Virgilian phrase in Kepler's account of the new star of 1604, see Boner, 2008b.

celestial refuse. We may be certain that these new forms exemplified the same fundamental design that underlay every other feature of Kepler's cosmology. While they have been overshadowed by his more memorable system of planetary harmony, their explanation presented an important challenge to his core principles. Their resolution calls for a closer look at the relation between beauty and truth in Kepler's world picture.[34] While my answer is incomplete, the emphasis I place on the archetypal principles found in Kepler's astrology casts new light on his cosmic vision. In Kepler's astrology, we come to know the relation between the celestial and sublunar spheres and how this spoke to "the deepest truths about the fabric of the world."[35]

What were the archetypal principles that underlay Kepler's astrology? The question of "which principles [were] archetypal and which [were] not" is central to every aspect of Kepler's thought.[36] Chapter 2 takes up this question in the context of Kepler's early career in astrology, 1594–1599. In the course of his correspondence with Johann Georg Herwart von Hohenburg (1553–1622) and other contemporaries, Kepler arrived at a select set of geometrical figures whose appearance in the heavens could elicit meteorological responses from the earth. Firmly convinced that the earth continued to enjoy "a place of privilege" in the new cosmology,[37] Kepler situated it at the center of an astrological symphony. The surrounding planets, including the sun and the moon, struck a chord in the earth by expressing archetypal principles in the form of particular configurations. These principles were metaphysical in nature and thought to reflect the mathematical essence of the human mind. In his natal astrology, Kepler argued that the arrangement of the heavens left a lasting impression on the newly born soul, serving as a source of reference for the rest of one's life.[38] At the moment of birth, Kepler would later claim, "the radiant harmony of the planets" first flowed into the vital faculty that fostered "the flame of life."[39] This moment was thus marked by an original configuration to which every other arrangement of the heavens later referred. As such, the heavens not only expressed archetypal principles in the form of

[34] On this relation, see also Jardine, 2009.
[35] Ibid., p. 276.
[36] Donahue, 2002, p. 297.
[37] Kozhamthadam, 1994, p. 175.
[38] On the original 'stamp' of the heavens on the human soul, see Rabin, 1987, p. 177: "In this way geometry becomes the archetype of the human being just as it is for the natural world."
[39] *JKGW*, 6, 278.30–32.

particular configurations, but their arrangement at birth left a mark much like a stamp or seal.

Kepler identified a soul in the earth that served "as an intermediary between the heavens and the sublunar world."[40] While he was not the only one to endow the earth with a soul,[41] Kepler equipped it with the ability to identify signs in the surrounding heavens and respond by intensifying the weather. The soul of the earth was also thought to express geometrical archetypes in nature, perpetuating the very principles employed in the original act of creation. Chapter 3 explores Kepler's conception of this soul and the part it played in his account of the new star of 1604. While he is widely known for the physical unification of the celestial and sublunar spheres,[42] Kepler relied heavily on the soul of the earth to account for change in the heavens. In his account of the new luminary of 1604, *On the New Star* (1606), Kepler envisioned it originating in the same way that some species of plants and animals were born from the earth. Kepler suggested that the celestial and sublunar spheres shared a creative capacity that produced new forms from decaying matter according to archetypal principles. From the fabrication of flora and fauna to the generation of new stars, this creative capacity served to bridge the heavens and earth. Geometry and mechanism were not the only unifying elements in Kepler's cosmology. His synthesis was woven from the fabric of things.

Chapter 4 looks further into the creative capacity of the cosmos in Kepler's account of the comets of 1607 and 1618.[43] While Kepler's claim that comets moved in straight lines has been rightly attributed to "the power of his archetypal principles,"[44] we know little about how he believed those principles were actually embodied. Kepler viewed the appearance of a comet or new star as a rare event, "in no way reducible to the fortuitous encounter of astronomical and physical effects."[45] At the same time, Kepler insisted that such celestial novelties were the consequence of natural causes. Originating from a condensed area of the ether and pursuing

40 Simon, 1979, p. 460.

41 On David Origanus (1558–1628) and his vitalistic view of the earth, see Omodeo, 2011, pp. 446–447.

42 Simon, 1979, p. 41: "The astronomical revolution entails the material unification of the different regions of the world, and astrology, like any other form of knowledge, cannot be based on celestial properties that would not have any guarantor on earth."

43 On the evolution of Kepler's cometary theory, see Barker, 1993, pp. 19–20. For a summary of his mature view of comets, see Drake, 1960, pp. 347–348.

44 Ruffner, 1971, p. 180.

45 Simon, 1979, p. 64.

a rectilinear path,[46] a comet was a natural phenomenon whose motion expressed some of the same archetypal principles that Kepler witnessed in other events. Although he supposed that comets could serve to admonish humanity, Kepler claimed that their formal and physical constitution was no different from any other natural thing. Neither the unpredictable appearance nor the obscure astrological meaning of a comet deterred Kepler from defining it as fully knowable. Chapter 4 closes with a discussion of the natural causes Kepler assigned to comets and their role in representing divine providence. I explore this role in relation to his greater endeavor to know "the complete divine plan of creation."[47]

Kepler's earlier astrological ideas "found their fully organized expression" in his cosmological masterpiece, *The Harmony of the World* (1619). There, Kepler reconsidered the configurations of the heavens and their archetypal resonance in the sublunar sphere. What line of reasoning lay behind this mature view of astrology, and how did it hold to his earlier idea of the earth as a living being? Chapter 5 discusses Kepler's defense of his account in a fierce dispute with Robert Fludd (1574–1637). Fludd attacked Kepler for his conception of a sublunar soul, together with what he called his "false astronomy" and "imaginary harmony."[48] Kepler's polemic with Fludd has often been portrayed as a clash between qualitative and quantitative cultures.[49] However, Kepler suggested some similarities he shared with his adversary. While he condemned Fludd's world view as "more poetic or oratorical than philosophical or mathematical,"[50] Kepler compared the animate core of Fludd's cosmology to his own notion of a universal faculty responsible for the continual creation of new forms. Fludd may have objected to Kepler's distinction of souls according to their different faculties, opposing "the application of the concept 'part,' "[51] but Kepler agreed that the creative capacity of the cosmos could not be confined to a single place. This faculty acted throughout the cosmos, Kepler claimed, and it continued to realize the original rudiments of creation.

---

46 On earlier views of comets as the consequence of ethereal condensation see Granada, 2012, p. lxxxix.
47 Bucciantini, 2007, p. 315.
48 *JKGW*, 6, 401.30–31.
49 See, for example, Hallyn, 1990, pp. 167, 248, 251–252.
50 *JKGW*, 6, 374.37–39.
51 Pauli, 1948, p. 199.

The idea of a sublunar soul may strike us today as "a stroke of superstition, an imaginative outburst, or even a pathological symptom."[52] It was certainly an idea that held sway for many early modern scholars who faithfully accepted the magnetic philosophy of William Gilbert (1544–1603). For Gilbert, the ability of the earth to revolve daily on its axis suggested the "animate nature" of our planet.[53] The earth was thus the seat of a magnetic soul and the body of "a giant animate creature."[54] Kepler and other contemporaries who accepted the magnetic philosophy measured the motion of the earth in the same empirical spirit as Gilbert. In the case of Kepler, he exacted a similar level of accuracy from his account of celestial change on the basis of earth-bound knowledge. Kepler remained deeply interested in the celestial novelties of his day and how they strengthened his campaign for a superior cosmology. In his attempt to unravel the full mystery first glimpsed by Copernicus, Kepler deployed his knowledge of these novelties doggedly. They belonged to the same series of problems that Kepler set out to solve in his system of celestial physics. By accounting for change, the new world was made whole.

That new stars and comets "employed the service of nature" in the same way as the earth did not lead Kepler to look beyond God.[55] While this book does not deal with the details of his theology, there is no denying the role theology played in Kepler's physical reasoning. Kepler saw the expression of divine order in the spontaneous generation of new forms. Sudden and unpredictable, these forms affirmed the narrative of new life that continued to unfold after the original act of creation. "In the beginning," Kepler wrote, God had begun the "perpetual process" of new life that nature would continue to fulfil.[56] Nature was thus given the great task of bringing the divine plan to fruition. "If God should tell nature to bring forth," Kepler wrote, "nature does that very thing."[57] This book considers how nature was thought to fulfill this function far beyond the earth, where Kepler well knew that "no one could be present."[58] It also examines the archetypal principles that Kepler identified in these new forms and

---

52 Simon, 1979, p. 177.
53 Henry, 2001, p. 116.
54 Ibid.
55 *JKGW*, 1, 291.27–30.
56 *JKGW*, 4, 65.1–2.
57 *JKGW*, 1, 291.30–32.
58 Ibid., 245.32.

their expression of the very "essence of God."[59] Made in the divine image, man was bestowed with the same principles from birth and re-awakened to their reality in the world of sensible things.[60]

Kepler sought understanding in the sensible world by way of a keen mind and creative intellect. He explored nature "with an incessant curiosity" and reflected on eternal questions of causation and conception.[61] Kepler has been called a spiritual child of Ptolemy and situated among the early modern successors of Plato and the Pythagoreans.[62] There is good reason for this view and we now enjoy a greater understanding of the mathematics that made up his metaphysical principles. The result affords a stronger sense of the "empiricism, mathematics, and metaphysics" that lay at the foundation of his most enduring discoveries in astronomy and optics.[63] This book moves beyond these studies, however, and attempts to complete our view of Kepler's cosmology. It explores the metaphysics that Kepler had in mind and how he thought it found expression in the physical world. The universal presence of these principles lay at the heart of his astrology and suggested a revolutionary synthesis. In turn, the fundamental unity of the world made the heavens the new target of natural philosophy. But the projection of natural knowledge beyond the sublunar sphere did not suggest a modern mechanistic synthesis. In the end, Kepler did not simply supply new and enduring elements of the mechanical philosophy, but defied the limits of a world reduced merely to matter and motion.

---

59 *JKGW*, 6, 55.11–20.

60 Bialas, 2004, p. 54: "The relation of the divine and human image finds expression in the immanent comprehension of geometrical things."

61 Gingerich, 2010, p. 10.

62 See Field, 1988, p. 190.

63 Bialas, 2004, p. 54.

CHAPTER ONE

KEPLER'S VITALISTIC VIEW OF THE HEAVENS: SOME PRELIMINARY REMARKS

Historians often view Johannes Kepler as an early forerunner of the mechanical philosophy. In their surveys, Kepler signals a new age in astronomy, recasting the heavens as a 'celestial machine' whose parts are driven by purely physical forces. The result is a revolutionary fusion of mathematical astronomy and physics, an interdisciplinary vision that draws on various areas to establish "an integrated system to explain the universe."[1] One element of this enterprise is what historians regard as the introduction of mathematically measurable forces in place of the spiritual forms of agency of earlier astronomy. For many, this represents a "radical revulsion of thought" which replaced spiritual principles with a form of mathematical physics that would resonate with the later theories of Descartes and Newton.[2] This elegant but simplistic view of Kepler deeply distorts his celestial physics, however, and obscures the complexity of early modern science. In this book, I argue that Kepler continued to accept the possibility of vitalistic principles in the heavens long after he introduced his system of celestial physics in the *New Astronomy* (1609).[3] I further suggest that these principles were an important part of his view of physics, which encompassed a wide range of explanatory resources that we might today call biology. By drawing on biology, Kepler extended earthly knowledge beyond the sublunar sphere, abandoning the ancient earth/heavens boundary that no longer held sway in the new cosmology.

I begin my study by reconsidering Kepler's use of mechanical metaphors and their meaning for his wider world picture. My focus on a familiar example shows that in at least one case Kepler did not deploy this metaphor beyond the scope of planetary motion, revealing the limited range of his mechanical imagery. Next, I turn to another subject of Kepler's celestial physics, novelty in the form of new stars. As I shall show, Kepler

[1] Osler, 2010, p. 60.

[2] Dijksterhuis, 1961, p. 310.

[3] I identify 'vitalistic principles' as the lower faculties of living beings. On the "noble and less noble faculties" in Kepler's early conception of the soul, see Escobar, 2008, pp. 19–21, 37.

accounted for new stars in a way that drew on biology rather than the more familiar laws of his physical astronomy. Less certain though no less significant for his causal account of the heavens, biology provided Kepler with a way of making sense of celestial mutability. At the same time, it reinforced the role of the earth as a basis for knowing the heavens. Just as Kepler projected magnetic powers among the planets, portraying them as "a pilot in a river" of magnetic current,[4] he conceived of celestial change in an earthly way. Kepler's use of more familiar knowledge was not merely convenient, however, for it spoke to an essential continuity between the heavens and earth. Although he allowed for a certain degree of ambiguity between the two, Kepler found in this application an effective way of explaining celestial events causally. In my final analysis, I explore a form of cosmic continuity more fundamental for Kepler than his 'celestial biology.' In his vitalistic view of astrology, Kepler identified principles of influence that exemplified the very essence of his cosmological view.

Kepler's Clockwork Metaphor

Perhaps no single passage from Kepler's works appears more often in historical surveys than his clockwork metaphor. In a letter to his patron, Johann Georg Herwart von Hohenburg, Kepler famously compared the causes of planetary motion to the gears and levers of a clock. "Nearly every variety of motion," Kepler wrote, "stems from a single magnetic corporeal force [*vi*], in the same way that all of the motions in a clock stem from a single weight."[5] Chiefly concerned with the physical causes of planetary motion, Kepler appears to contrast his conception of the "celestial machine" with the Platonic view of the world soul.[6] In the *Timaeus*, Plato had suggested a soul as the causal essence of the cosmos, a divine living being synonymous with "a single spherical universe in circular motion."[7] Aristotle had criticized Plato's world soul for producing the motions of the heavens by a perpetual form of exertion that "could not possibly be painless or blessed."[8] For Aristotle, the idea of a soul continually sapped by the sustenance of motion was anything but divine. For his own part, Kepler

[4] *JKGW*, 3, 349.11.
[5] *JKGW*, 15, no. 325, 59–61.
[6] Ibid., 57.
[7] Plato, 1977, p. 46.
[8] Aristotle, 2006, p. 135.

denounced the world soul as a way of diverting the glory of the divine creator. Such a soul would, in effect, steal "the glory of the artificer [*gloria artificis*]."[9] Rather than rob supreme regard for the creator by accepting an entirely self-sufficient soul, Kepler argued for a system of planetary motion made up of secondary causes. Siding in this way with Aristotle, Kepler gauged his system against the highest standards of Euclidean geometry. Drawing on the explanatory resources of natural philosophy and the analytical tools of mathematics, Kepler accounted for the paths of the planets according to a new system of celestial physics.[10]

For many historians, Kepler's metaphor marks the rising popularity of the clock as a resource for representing a new and revolutionary view of nature.[11] According to this view, the supreme order and regularity of God's work could, on a smaller scale, be compared with the mechanical craft of the clockmaker. Thus, the clock provided a ready reference for the many regular motions occurring in nature as well as human society.[12] Forcefully expressed by the natural philosophy of Descartes, this view suggested that the motions of the heavens could be understood according to the same physical principles as machines. In the *Discourse on Method* (1637), Descartes endeavored to explain even the biological processes of the human body in terms of mechanical principles. Such a synthesis of the natural and artificial worlds "counted as a violation of one of the most basic distinctions of Aristotelian philosophy."[13] At the same time, it replaced the scholastic properties of purpose and sentience with attributes that were expressible in terms of shape, size, and motion.[14] In her survey of "the transition to the machine as the dominant metaphor binding together the

---

9 *JKGW*, 15, no. 325, 58–59.

10 Ibid., 61–67: "And I take up this physical reasoning [in the *New Astronomy*] according to the numbers and geometry, in such a way that I do not frighten you with the fantasies of Alpetragius, who attempted before Fracastoro to effect everything by concentric spheres. Alpetragius did not attend to the numbers, however, or else he would have held on [to eccentric spheres] and realized that he was dreaming. Tycho rejected the orbs, and now I show how the planets may move without the orbs and how eccentricity may occur, etc."

11 At the same time, the use of the clock metaphor by early modern scholars such as Kepler suggests "the increasing importance of the mechanical arts" in the period; see Neumann, 2010, p. 138.

12 On the different uses of this metaphor "for the cosmos, society, and natural philosophy" in Continental Europe and Great Britain, see Henry, 2008, pp. 106–107.

13 Shapin, 1996, p. 30. Cf. Mayr, 1989, p. 55. On the role of alchemy in the shifting boundaries between art and nature in medieval and early modern Europe, see Newman, 2006.

14 Maier, 1938, p. 1: "With few exceptions, philosophers agreed that the scholastic forms and qualities belonged to the past and that everything qualitative in the world could be explained mechanically, that is, reduced to shape, size, and motion."

cosmos," Carolyn Merchant considers this "new machine-like world" to be responsible for literally removing life from nature.[15] Merchant suggests that Kepler's metaphor serves as a clear sign of "the failing plausibility of the organic model of the macrocosm."[16] For Merchant and many others, Kepler's famous metaphor represents a precursor of the mechanical world picture, an early glimpse of the cold "world of quantity" which Alexandre Koyré so famously described in his account of the Newtonian synthesis.[17]

Recent surveys of the role of the mechanical philosophy in seventeenth-century science continue to reflect the studies of Anneliese Maier and Eduard Dijksterhuis. In their books, both titled *The Mechanization of the World Picture*, Maier (1938) and Dijksterhuis (1961) see the mathematization and the mechanization of nature as the distinguishing features of early modern science. For Maier, the shift from the speculative philosophy of the scholastics to the mathematical views of Galileo and Gassendi was accompanied by the appropriation of the ancient atomic philosophy. Significantly, Maier suggests Gilbert's *On the Magnet* (1600) as "perhaps the strongest precursor of the general reception of atomic thought" in the seventeenth century.[18] In this way, Gilbert serves as a counterpart to Harvey, who similarly put forward an animistic theory subsequently deployed by others in support of mechanical views.[19] Maier does not, however, credit Kepler with applying Gilbert's magnetic theory to his system of celestial physics. Rather, Maier briefly recognizes Kepler for his quest to uncover a universal mathematical harmony.[20] Dijksterhuis, on the other hand, focuses more on Kepler's dynamic study of the heavens, which displays "the direction in which science as a whole was to evolve."[21] Although Dijksterhuis observes a certain degree of caution in his characterization, Kepler represents for him "a radical revulsion of thought."[22]

---

15 Merchant, 1989, pp. xxii, 20.

16 Ibid., p. 128.

17 Koyré, 1965, p. 23.

18 Maier, 1938, p. 11. Cf. Hooykaas, 1972, p. 62. On Maier's view of late scholasticism as "a precursor of early modern physics," see Cohen, 1994, pp. 58–59.

19 Maier, 1938, p. 12: "In later years, Harvey's discovery of the circulation of the blood became particularly significant. It seemed to open the way for a mechanical explanation of the very functions of life and in this way sanction corresponding efforts in the inorganic field."

20 Cf. ibid.: "In his conception of the qualities and sense perception, however, Kepler continues to follow the traditional teaching. In this respect, there is scarcely any room for regarding him as a forerunner or predecessor of Descartes."

21 Dijksterhuis, 1961, p. 313.

22 Ibid., p. 310.

Kepler's famous metaphor comparing the cosmos to a clock is presented as the "first tentative steps" of a process that would culminate in the celestial mechanics of Newton.[23] Dijksterhuis has little to say, however, about Kepler's theory of an ensouled earth,[24] and even less about the vitalistic speculations which Kepler entertained long after his letter to Herwart. It would seem that Dijksterhuis' "radical revulsion of thought"—the rejection of "an animistic in favor of a mechanistic conception"—reflects only a part of Kepler's overall world picture.[25]

Partial views of Kepler's cosmology affirm still earlier studies by scholars such as Whewell, in whose "resounding success story" of science Kepler figures as a founder of the mechanical philosophy.[26] Focusing primarily on Kepler's laws of planetary motion, Whewell pieced together only parts of Kepler's cosmic perspective without considering his view as a whole. For many, painting a complete portrait of Kepler would amount to a sort of chronological schizophrenia. Areas of interest in which Kepler did not enforce the same standards of mechanical precision and observational accuracy would come to be seen as the remnants of Renaissance philosophy. In Koyré's memorable account, Kepler became "a veritable Janus," whose one face "looked back to the hierarchically ordered, human-centered medieval cosmos," while the other "looked forward to the universe uniformly regulated by general mathematical laws."[27] In part, this paradoxical view is the product of what Bruce Stephenson refers to as the problem in the poem "The Blind Men and the Elephant." Like the elephant in the poem, Kepler's overall enterprise—covering the areas of astronomy, natural philosophy, optics, and theology, among others—"has a very different feel depending upon which part of it one seizes on."[28] Those who have chosen to focus on the more familiar parts of astronomy, mathematics, and optics have often been forced to allow space for subjects such as astrology, which, as Koestler would quip, "constantly intrudes even in

23 Ibid., p. 314.

24 Curiously, Dijksterhuis claims that Kepler abandoned this theory in the *New Astronomy* (1609). On Kepler's conception of the sublunar soul in Book 4 of the *Harmony of the World* (1619), see Schwaetzer, 1997, pp. 222–239.

25 Dijksterhuis, 1961, p. 310.

26 Cohen, 1994, pp. 36–38. According to Whewell, Kepler's main focus as "a physical investigator" was "to discover a mathematical rule" to account for the observed facts and to find the laws of planetary motion that would "conform to the same conception of causation." See Whewell, 1847, p. 384.

27 Koyré, 1992, p. 120; Jardine, 2000, p. 363.

28 Stephenson, 1994, p. ix.

Kepler's classic scientific works."[29] In the following chapters, I take up Kepler's astrology as a way of illustrating many of the core principles of his cosmology. By seizing on Kepler's theory of the sublunar soul and his speculations on comets and new stars, I hope to shed further light on familiar parts by examining features seldom seen.

Before proceeding directly to Kepler's astrology, however, let us take a closer look at his clockwork metaphor. What else might Kepler have meant by this expression, and how well does it capture the mechanical motif of early modern science? To begin with, it is quite possible that Kepler was aware of earlier mechanical metaphors whose ancient and medieval authors appear less often in our surveys of the Scientific Revolution. These earlier expressions resemble Kepler's own idea of a "celestial machine" by suggesting a series of causes that explain natural phenomena systematically.[30] In his epic expression of Epicurean philosophy, the Roman author Lucretius had written of a "machine of the world [*machina mundi*]," a "congress of matter" that would fall apart after many years.[31] Beginning with the "earth, sea, sky, and stars," this world machine had produced a series of parts, including "the sun and the globe of the moon," as well as "the living things that arose from the earth."[32] It was neither free will nor "some plan of the gods" that guided "the course of the sun and the motion of the moon," but rather a governing "force [*vis*]" that could be grasped physically.[33] Although we may read Lucretius as representing the motions of the heavens mechanically, it should be kept in mind that he also compared the cosmos to "a mortal body" that would eventually expire "with a frightful crash."[34] Such a dynamic conception accounted for the cosmic state of complexity in terms of the chance interaction of material particles. Just as the "congress of matter" had led to ever greater levels of complexity, so would it one day give way to destruction.

Closer in Christian spirit to Kepler's clockwork metaphor were the mechanical views of some medieval scholars.[35] In the final lines of his

[29] Koestler, 1960, p. 39.

[30] On later theories in natural theology of a world structure whose "elaborate clockwork" suggested the signature of a "technically and mathematically well versed designer," see Neumann, 2010, pp. 134–135.

[31] Lucretius, 1924, pp. 384–386. Cf. Mayr, 1989, p. 39: "To compare the world with a clock was only an extension of the ancient custom of calling the world a machine."

[32] Ibid., p. 384.

[33] Ibid.

[34] Ibid., pp. 384–386.

[35] On Kepler's burning opposition to Epicurean philosophy as "an affront to God Almighty," see Boner, 2007. Apparent resistance by some theological authorities to the

famous work on spherical astronomy that appeared in the early thirteenth century, Sacrobosco referred to the words of Dionysius the Areopagite on the miraculous nature of the solar eclipse that took place during Christ's Passion. "This eclipse was not natural," Sacrobosco wrote, "since a solar eclipse should occur at new moon or round about then."[36] Confirming the supernatural occurrence of the eclipse near full moon, Sacrobosco reported that, "at the time of Christ's Passion, Dionysius the Areopagite said, 'Either the God of nature suffers or the machine of the world [*mundi machina*] is dissolved.'"[37] Sacrobosco did not question "the God of nature" any more than Dionysius. Instead, he suggested a supernatural break in the systematic course of celestial motion. The metaphor of the world machine conveyed an "unseen rationality,"[38] a system of order and regularity that displayed the causal links of divine providence. Kepler had certainly studied Sacrobosco at the University of Tübingen, where he would have found an edition featuring a preface by Philipp Melanchthon (1497–1560). The famous Lutheran reformer had told his readers that astronomy "taught about Providence,"[39] and that our eyes were given to us "especially for guiding our search for some knowledge of God."[40] Melanchthon proudly pointed to the recent revival of astronomy, acknowledging the part played by Sacrobosco as well as the role of his native Germany.[41] "After several centuries of disgrace," Melanchthon wrote, astronomy had "recently begun to flourish in Germany," where it had been "restored by Peurbach

---

intelligibility and predictability of celestial motion led Philip Melanchthon to label them as Epicureans in his preface to Sacrobosco's *Sphere*; see Sacrobosco, 1543, 4v–5r: "There are some Epicurean theologians who mock the whole of astronomy. They not only deny their faith in predictions, but they disparage any knowledge of the [celestial] motions. Let us allow them to act foolishly along with Epicurus, for they are in a condition that would require medical doctors rather than geometers. It is a flagrant form of madness to reject knowledge of the [celestial] motions, which has the finest and firmest foundations, in such a way that we may omit that other, predictive part entirely." Cf. Kusukawa, 1995, pp. 128–129. As my reference to Lucretius reveals, I do not anticipate a single thread of thematic continuity among mechanical metaphors any more than Mayr does in his "large collection of clock metaphors." See Mayr, 1989, p. 30.

[36] Sacrobosco, 1543, 30v. On the impossibility of the eclipse at Passover, see Gingerich, 2004, p. 191.

[37] Sacrobosco, 1543, 30v.

[38] North, 2005, p. 201.

[39] Kusukawa, 1995, pp. 129, 131.

[40] Sacrobosco, 1543, 3v.

[41] Ibid., 2v: "We see that very few works survive the test of time, especially in the schools, where judgment is rather hard-nosed. Yet [Sacrobosco's] book is read now for many centuries with the utmost approval by all of the learned."

and Regiomontanus."[42] The efforts of these two authorities had brought a better and broader understanding to astronomy. With it, the order and regularity of the heavens had firmly convinced him of the reality of astrological influence.[43]

With the invention of the mechanical clock around the end of the thirteenth century, Nicole Oresme (ca. 1320–1382) and others saw a new way of representing the motions of the heavens mechanically. In one account, Oresme compared the system of celestial motion to "a material clock," whose many "motions and wheels [were] as nearly commensurable as possible."[44] Oresme deployed this metaphor in defense of the arithmetical arrangement of the heavens, in support of celestial commensurability. If the heavens displayed such a high degree of rationality and regularity, Oresme wrote, "how much more ought we to think of that architect who, it is said, made all things in number, weight, and measure?"[45] Oresme also expressed a familiarity with apparatus such as the "clock of the heavenly motions," an instrument which could apparently produce the positions of the planets and the fixed stars "at any moment of the day or night."[46] Created for the practice of astrological prediction, this clock was made up of countless wheels so cunningly constructed that all of their motions were governed by a single counterbalance. We are told by Oresme that, among other "astrological machines," the clock of the heavenly motions attracted the attention of all of the learned community "in astronomy, philosophy, and medicine."[47] Such devices were increasingly found in the

---

[42] Melanchthon, 2003, p. 166: "And students of this subject ought to be moved as well by the success of the country, since astronomy, after falling into disgrace for several centuries, has recently begun to flourish in Germany, [where] it has been restored by the two highly gifted men, Peurbach and Regiomontanus, the former from Upper Austria and the latter from the vicinity of Franconia. History testifies that these heroes were divinely inspired by some singular force to illuminate these arts." Cf. Kusukawa, 1995, p. 129.

[43] Sacrobosco, 1543, 7v–8r: "Since this admirable order and disposition [of the heavens], along with the most certain laws of celestial motion, show that this work [*opificium*] consists in supreme reason, it cannot follow that the stars possess neither signification nor effect."

[44] Grant, 1971, pp. 294–295.

[45] Ibid. On Oresme's use of clock imagery in his commentary on Aristotle's *On the Heavens* (1377), see Mayr, 1989, pp. 38–40. "Parallel to the development of the mechanical clock," Neumann notes a "growing iconographical connection" between the clock and "*temperantia*, the *virtus* of mass and measure," along with divine knowledge of "the order of the world according to number, weight, and measure in the Book of Wisdom." See Neumann, 2010, p. 130.

[46] Coopland, 1952, pp. 15–16.

[47] Ibid. On Oresme's reduction of astrology to "light and motion in the heavens and the action of the four primary qualities here below," see Thorndike, 1934, p. 414.

possession of patrons across Europe, as interest in mechanical marvels spread rapidly.[48]

Kepler acquired a close familiarity with astronomical apparatus at the court of Holy Roman Emperor Rudolph II (1552–1612). There, Kepler witnessed the work of the Imperial Instrument Maker, Jost Bürgi (1552–1632). One of Europe's premier clockmakers, Bürgi had created a series of celestial spheres and other complex clocks at the court of his previous patron, Wilhem IV (1532–1592), Landgrave of Hesse-Kassel. In 1592, Bürgi had delivered to the imperial court a celestial sphere showing the motions of the planets and the fixed stars.[49] Having left a lasting impression, Bürgi returned permanently to Prague in 1603 and quickly earned the admiration of Kepler. At his workshop in the imperial residence, Bürgi oversaw the construction of apparatus whose complexity required him to prepare his designs in perspective rather than in the planar form of his peers.[50] Kepler praised Bürgi for combining this mathematical vision with an extraordinary mechanical ability, a rare set of skills which Kepler thought would one day win Bürgi as much fame in his own field as Dürer enjoyed

---

48 As Mayr suggests, the early mechanical clock often served as a source of artistic expression and entertainment rather than as an instrument of accurate timekeeping. See Mayr, 1980, p. 1.

49 Maurice, 1980, p. 89. In the course of his report on the star that appeared in the constellation of the Swan in 1600, Kepler summoned Bürgi as a witness to "the novelty of this star." While at Kassel, Bürgi had engraved a celestial globe as part of a planetarium which would be given as a gift to the Emperor. Soon after the appearance of the star, Bürgi found himself at the imperial court, where he could confirm the absence of the star from the globe. See *JKGW*, 1, 307.32–308.9: "When several errors were discovered in Ptolemy's tables and many more glaring things were also found missing from Ptolemy beyond the forms of the stars, the Landgrave asked for a mechanician [*mechanicus*] to engrave everything so that it would be quickly brought into conformity with the heavens. Bürgi fulfilled this task diligently, producing a perfect globe. A gift fit for royalty, it was sent by Landgrave Wilhelm to Emperor Rudolph and remains to this day among His Majesty's treasures ... Shortly after [the appearance of the star], Bürgi came to the imperial court, where he devoted his attention to repairing many wonderful works, among them also that globe which he had previously produced. This globe showed no sign of anything except that ancient star given by Ptolemy in the breast of the Swan. Thus, Bürgi speaks to the novelty of this star with complete confidence."

50 Maurice, 1980, p. 88. In a letter to Tycho, Wilhelm described Bürgi as "a second Archimedes." See Caspar, 1993, p. 165. Archimedes was celebrated by such authors as Cicero for his model "imitating the revolutions of the [celestial] sphere." See Cicero, 2005, pp. 208–209. The Syracusan was also praised for putting his model in perpetual motion, a predecessor of the "miraculous discovery of perpetual motion" made by Cornelis Drebbel (1573–1633) in the published works of "Drebbel's most ardent supporter," Gerrit Schagen. See Vermij, 2002, pp. 92–95.

in painting.[51] Kepler collaborated with Bürgi on mechanical projects such as the water pump,[52] and his presence at the imperial court promised a continual form of contact with the artisans working under the Imperial Instrument Maker. Kepler made use of this mechanical knowledge as an analogical resource in the *New Astronomy* (1609),[53] whose account of planetary motion he previewed with his clockwork metaphor. Yet how comprehensively did Kepler apply this analogy? Closer analysis of his astrology shall show that Kepler meant his comparison for the motions of the planets rather than for the cosmos as a whole.

A contemporary of Oresme who similarly compared the cosmos to a clock was Heinrich von Langenstein (1325–1397). In a series of lectures on the Book of Genesis, Langenstein traced the sequence of efficient causes that moved the universe "from God, through the stars, and into the elements."[54] According to Langenstein's discussion of the fourth day, God created the cosmos in a series of parts and, once all of the parts were in place, put them in motion. Langenstein compared this process to the creative act of a clockmaker, "who fashioned his masterpiece bit by bit," until all of the parts were arranged in the right place and the clock could begin running.[55] Langenstein also looked to this metaphor as a way of making sense of astrology. Like the transfer of motion in a clock, the link between the heavens and the earth was limited to "a simple mechanical process."[56] Since the stars were created to serve as signs, Langenstein sug-

---

[51] *JKGW*, 1, 307.14–19: "Jost Bürgi is the Imperial Instrument Maker [*Automatopoeus*]. Although Bürgi is without languages, he easily surpasses many professors of mathematics in his knowledge and study of mathematical matters. As a result, he possesses a practical skill so peculiar, in fact, that a future age will celebrate him as a leader in this field as much as they celebrate Dürer in painting, whose fame grows as a tree with the hidden passage of time." The final clause of this passage is taken from Horace's ode to Marcellus (*Odes*, 1:12, l. 45).

[52] On Kepler's development of a water pump, see Di Liscia, 2009, 667–691.

[53] On Kepler's account of planetary motion "according to the law of the lever," see Martens, 2000, p. 107.

[54] Steneck, 1976, p. 92. Langenstein gave these lectures at the University of Vienna, where he taught theology from 1385 until his death in 1397.

[55] Ibid., pp. 92, 149: "[I]n the final analysis . . . Henry's world is mechanical. There is no other way to view a world that acts through a chain of events with the superior governing the inferior. The world is a *machina mundi*, a giant clock."

[56] Ibid., p. 94. In his preface to Sacrobosco's *Sphere*, Melanchthon similarly suggested the transfer of motion as the efficient cause of astrological influence. See Melanchthon, 1543, 5r: "And so I judge Aristotle to have spoken correctly when he said that this inferior world is governed by that superior one and that superior bodies are the cause of motion in inferior ones. And he prudently supplies a foundation [*rationem*] for this, namely that

gested, their association with the earth was mainly a mechanical transfer of motion.[57] The complexities of this process were so great, however, that Langenstein held out little hope of understanding the causal source of celestial influence. Human beings, he argued, could not climb the chain of causality to the heights of the heavens. Their progress was prevented by the elevation of explanations "known only to God."[58] As we shall see, Kepler did not share the mechanical view of Langenstein, whose criticism of astrology coincided in many ways with that of Oresme, his colleague at the University of Paris.[59] For Kepler, astrology also involved vitalistic principles that made the influence of the heavens possible.

Over the course of the seventeenth century, the metaphor of the clock grew with "unqualified approval" as a symbol of order and regularity in the civil sphere.[60] Scholars turned to it as a way of expressing "the submission of inner savagery" as a source of personal and political order.[61] Johannes Amos Comenius (1592–1670) applied the imagery of the clock to the art of educating society and the soul in his new system of universal education, *The Great Didactic* (1657). "Man, like the universe," Comenius wrote, "resembles a mighty clockwork made up of bells and wheels so elaborately that all of the parts of the work interlock harmoniously and progress perpetually."[62] Comenius saw nature as ordered in such a way that it served as an example for the order achievable by the right form of education.[63] There was no essential difference for Comenius between "the idyllic image of the organic growth of the student" and the mechanical method that he suggested for the student's learning.[64] Education was now

since the origin of motion is from the heavens, it follows that celestial motion is the cause of motion in everything else."

57 Langenstein argued that the celestial orbs created "a barrier to the movement of the elements from the inferior to the heavenly region." We may assume that he also thought they prevented the elements from moving in the opposite direction. On the elemental essence of the stars, see Steneck, 1976, pp. 61–62.

58 Ibid., p. 66. Cf. Thorndike, 1934, p. 479.

59 Thorndike characterizes Langenstein's efforts in astrology as a reinforcement of Oresme "by more specific and limited criticisms of particular parts of astrology." See Thorndike, 1934, p. 492.

60 Mayr, 1989, p. 41.

61 Sutter, 1988, p. 15.

62 Comenius, 1954, p. 42: "And man is nothing other than harmony with respect to his body and soul."

63 Ibid., p. 78. On Comenius's "strongly vitalistic view of the universe," see Giglioni, 1995, pp. 27–33.

64 Sutter, 1988, p. 20. Comenius envisioned the end of education as a state similar to that before the Fall.

an exercise that aspired to nature, as boundaries between art and the once inimitable order of nature broke down. The synthesis of art and nature has been seen as a sign of "the triumph of the mechanistic world picture" in the seventeenth century.[65] In the case of Comenius, the analogy of the clock extended to the very constitution of the soul, where everything from the will to the passions found a clock-like counterpart:

> Analogically, the will is the main wheel in the motions of the soul. The weights, which impel the motions and incline the will here and there, are the desires and passions. The anchor that enables and inhibits the motions is reason, which measures and determines where and how much one must aim for or avoid certain things.[66]

Comenius went further than Ficino (who compared the soul in the body to "the ordering principle [*temperatio*]" in a clock)[67] by breaking down the soul into parts and comparing them to particular components of a clock. Comenius continued to share with Ficino, however, the view that the soul acted according to "a natural art [*ars naturalis*]."[68] When the soul's activity was properly ordered by this art, a perfect example was made for the simulation of nature by the mechanical measure of time. Comenius went so far as to model the soul on nature, claiming that nature followed a course "with greater precision than a living body led by a soul."[69] Moving from the soul to society and on to "the unchangeable order" of the heavens,[70] Comenius presented a path that led to ever higher levels of mechanical harmony.

---

65 Hooykaas, 1972, p. 61.

66 Sutter, 1988, p. 17.

67 Neumann, 2010, p. 150: "This citation from Book 6 of Marsilio Ficino's magnum opus, *Theologia Platonica de immortalitate animorum* (Florence, 1482) is the one and only passage in which Ficino makes use of the clock as a model of comparison for describing the relationship of bodies and souls."

68 Ibid., pp. 150–151.

69 Comenius, 1954, p. 77: "Yet what hidden force brings this to pass? None other than the apparent rule of order everywhere. This is the force of the proper arrangement of all of the parts according to the right number, measure, and order, with every part possessing the proper means to fulfill a prescribed function and thereby contributing to the final result of the whole... In this way, everything proceeds with greater precision than in a living body led by a soul. Yet if anything in it should break, crack, distort, slacken or split open, even the smallest wheel, the tiniest axle, or the finest screw, it will all stand still or depart from its proper function. Thus, it is obvious from this that everything depends on the order of the parts."

70 Cf. ibid., p. 75: "How does the course of time uncoil so firmly into fixed segments of years, months, and days without any error? By the one and unchangeable order of the firmament!"

Portrayed as a pioneer of the mechanical views of Comenius and his contemporaries, Kepler appears as an early forerunner of the mechanical philosophy. In this vein, his clockwork metaphor is viewed as the expression of a "typical transition figure," caught between the extremities of "scholastic prejudice" and the radical ideas of Isaac Beeckman (1588–1637) and other mechanically minded contemporaries.[71] Kepler is even said to have wavered back and forth between these two extremities without ever fully departing from the "traditional organic point of view."[72] It is hard to believe, however, that the same author who presented his polyhedral hypothesis in the *Cosmographical Mystery* (1596) as "an authentic model of the world" would develop a habit of "changing over from one view to the other."[73] Throughout his life, Kepler remained thoroughly convinced of his ability to arrive at "a true image [*Ebenbild*] of the creation,"[74] and we cannot interpret his metaphor as an expression meant "only intermittently."[75] In this book, I take up the challenge of showing how Kepler's celestial physics formed part of the same world picture that would embrace various vitalistic views over the course of his career. Apart from planetary

---

[71] Hooykaas, 1972, p. 62. Incidentally, Beeckman was an avid reader of Kepler and remarked on his work repeatedly. In his review of Kepler's *New Year's Gift* (1611), Beeckman objected to the idea of a 'formative principle' [*conformatrix*] at the heart of Kepler's account of the hexagonal shape of snowflakes. Such a principle was "ridiculous and unworthy of philosophy," Beeckman wrote, and served only to "conceal the cause rather than uncover it." In place of this principle, Beeckman put forward a purely mechanical explanation of the material properties of water particles. See Beeckman, 1945, p. 34. On Beeckman's plans to publish "a mechanistic interpretation of Kepler's theories" after studying his physical astronomy "intensively for some time," see Van Berkel, 1983, p. 133. Beeckman even lobbied the University of Leiden at one point to bring Kepler to Holland; see ibid., p. 127. I thank Rienk H. Vermij for pointing out the many connections between Beeckman and Kepler.

[72] Mayr, 1989, p. 61. Cf. Hooykaas, 1972, p. 62.

[73] Ibid., p. 62. Hooykaas associates Kepler's periodic return to a mechanistic perspective with his "renewed efforts to make mechanical models of the world." At one point when Hooykaas attributes an animistic perspective to Kepler, however, we find him preparing a model of his polyhedral hypothesis for the Duke of Württemberg, Friedrich I (1557–1608). Originally, Kepler envisioned this model as a large chalice whose multiple layers would measure the relative distances of the planets from the sun. Each of these layers would contain a different alcoholic beverage that would flow from "seven taps, covered by the images of the planets" in such a way that Kepler would guide his guests to the drink of their choice according to the properties of the relevant polyhedron. As Kepler explained to the Duke, "in this model Your Majesty will see why God made use of no more than five regular solids in this order, with each one situated between two orbs in such a way that it touches the outer one with all of its points and the inner one with all of its faces." See *JKGW*, 13, no. 30, 23–26. Although this model would not reproduce planetary motion mechanically, it coincided with Kepler's early speculations on the motive power of the sun.

[74] Hooykaas, 1972, p. 66.

[75] Mayr, 1989, p. 61.

motion, Kepler attempted to provide a causal account of other celestial events such as comets and new stars. In contrast with the constancy of the planets, these events were subject to change. To explain them, Kepler applied biological ideas to the ethereal body of the heavens. There, he suggested that decay and renewal took place among the stars in the same way as the cycle of life occurred on earth. That Kepler continued to entertain these ideas long after his clock analogy allows us to understand his later works as a development of his earlier ones rather than as a puzzling rejection. To be sure, Kepler did not suddenly cast aside these ideas only to return to them later.[76] Kepler's insistence on a kind of 'celestial biology' should not be viewed as a contradiction but as a central element in the evolution of his thinking.

Kepler was deeply concerned that his discovery of a celestial apparatus, if animate, would take away from "the glory of the artificer [*gloria artificis*]."[77] His concern with this question recalls the admonition of Augustine, who bid the onlooker "not to devote all his attention to the appearance of the material object that has been produced, but to look beyond it and recall with affection the one who produced it."[78] For many, of course, celestial motion was more than a material object. Marsilio Ficino (1433–1499) had identified it as the manifestation of "the most perfect animal" of the universe.[79] For his own part, Kepler opposed any sort of spiritual agency with his mechanical metaphor. More importantly, Kepler looked beyond "the beauty of every created thing" and, like Augustine, sought the model of the master craftsman.[80] An essential part of this process was to provide a causal account for any given phenomenon. In the case of planetary motion, Kepler suggested "a single magnetic corporeal force," which he considered the cause of the orbital motion and eccentricity of every planet. Such a level of causal economy inspired Kepler to call his system of celestial physics an "astronomy without hypotheses" that

---

76 According to Granada, Kepler's real turning point was redefining the relation of the disciplines, "loosening the ties between astronomy and theology... and seeking to unify astronomy and natural philosophy in a global consideration of the finite cosmos as a living organism endowed with a soul that gives rise to analogous productions at the different levels of reality." See Granada, 2009, p. 398.

77 *JKGW*, 15, no. 325, 58–59.

78 Augustine, 1993, p. 62.

79 Hirai, 2002, p. 274.

80 Augustine, 1993, p. 62. On the links between beauty and truth in Kepler's cosmology, see Jardine, 2009.

conformed to the very "laws of nature."[81] The simplicity of this system held a powerful place for Kepler, and it is no surprise that he approached more complex matters such as comets and new stars with greater hesitation. Regardless of his reluctance, however, Kepler did not neglect these novelties. In his causal account of the heavens, Kepler considered a much broader expanse of explanation, one that drew on the resources of every area of early modern physics [*physis*], including biology. That his clockwork metaphor has come to represent this explanation cuts Kepler's cosmology short and forces us to reconcile those parts of his world picture which have been found "alive and kicking."[82]

We gain a sense of the greater expanse of Kepler's celestial physics by looking more closely at his cosmological theory after the *New Astronomy*. In his study of the new luminary that appeared in the sky in 1604, *On the New Star* (1606), Kepler made clear that his mechanical metaphor did not spell the end of his speculations on spiritual agents. Kepler had begun this book in the spring of 1605, after the bulk of the *New Astronomy*—and with it, the core of his physical astronomy—was already complete.[83] Here, he considered as the efficient cause of the new luminary a faculty equally active in the heavens and on the earth. This faculty was more like "a spirit," Kepler said, which had produced the new luminary possibly "in the upper reaches of the ether in response to the appearance of the great conjunction [of 1604] on earth."[84] For this response to take place, Kepler required the presence of the spirit "pervading the entire world machine [*mundi machinam*]."[85] Are we to interpret this conjecture

81 *JKGW*, 15, no. 357, 51–58: "For what is it that carries the planets around the sun? (Even Tycho and Copernicus agree on this.) What is it, then, other than a magnetic effluvium [*effluvium magneticum*] of the sun? And what is it that makes the motions of the planets around the sun eccentric, forcing them to approach and recede from it? Why, a magnetic effluvium from the very bodies of the planets, according to the direction of their axis! And all of these lines of reasoning in the *New Astronomy* are joined together in such a way that it is necessary that either they all be false or common to the planets and their particular qualities."

82 Koestler, 1960, p. 42. For an example of this anachronism, look no further than my early study of Kepler's astrology (2005).

83 In a letter of 6 March 1605 to Michael Mästlin (1550–1631), Kepler claimed that the *New Astronomy* would contain "about sixty or seventy chapters," fifty-two of which were already written. See *JKGW*, 15, no. 335, 182–184. Cf. Granada, 2010, pp. 126–127.

84 *JKGW*, 1, 276.20–22.

85 Ibid. Later, Kepler wrote that he had allowed for the possibility of "such a universal spirit in my book," but not without also supposing a soul in the earth, whose existence he had proven through "many of its operations." See *JKGW*, 4, 143.35–37. In his *General Consultation* (1666), Comenius suggested in sympathy with Tommaso Campanella (1568–1639) that universal perception could be considered as a sensation. In his summary, Comenius

as a clumsy juxtaposition of mechanical apparatus and a spiritual motor? Kepler seems to suggest another story. By supposing a spirit, he did not break with his earlier conception of physical causes but borrowed from biology to make sense of a phenomenon that fell outside the scope of the *New Astronomy*. The new luminary had appeared and later vanished, a sign of celestial growth and decay, and Kepler explored the issue with a set of tools that simply came from another part of his causal inventory. As he explained in a letter of 1605, the universal presence of such a spirit would order all things, "providing new creatures with appropriate bodily apparatus."[86] Seen as a source of renewal, this spirit would create "comets and stars from the redundant matter of the ethereal region" in the same way that it spontaneously produced a wide variety of living species from the excesses of the earth.[87] In the end, Kepler's consideration of other causes to account for things beyond planetary motion was an extension of his effort to merge the heavens and earth under a single causal system.[88] All things were of a piece.

Kepler was not alone in his effort to explain celestial events by bridging earthly boundaries. A Copernican contemporary, Philip Lansberg (1561–1632), compared the heavens and earth to the parts of an egg, "with the sphere of the fixed stars as its shell."[89] Together, the two areas were provided with everything necessary to produce "a chick" representing the range of things that could be created in the cosmos. Although Lansberg

---

wrote that Campanella had conceived of "the whole world machine [*mundi machinam*]" as animate and sentient. See Giglioni, 1995, p. 26.

[86] *JKGW*, 15, no. 358, 746–751: "You may see that everywhere there is something moist that contains a seminal reason [*rationem seminariam*] and effects a variety of species in such a way that not just any caterpillar is born from the leaves of just any tree. Rather, every species of caterpillar comes from a particular species of tree. In fact, this spirit appears to ensure throughout the world that all things are mutually ordered, providing new creatures with appropriate bodily apparatus." Kepler's account of seminal reason here recalls that of Marsilio Ficino (1433–1499), who identified it as the efficient principle of form for every living species. Kepler explicitly compares his account, however, with that of Cornelius Gemma (1535–1578). For more on these "reason-principles" in Ficino's view of the vegetative power of the World Soul, see Hirai, 2002, pp. 271–279.

[87] Ibid. Among the earthly excesses that produced living species spontaneously were "the sweat of women into lice and the sweat of dogs into fleas, dew into caterpillars and grasshoppers, mud into eels, marshland into frogs, water into fishes, earth into plants, cadavers into worms, dung into beetles, and an infinite number of new and strange things."

[88] Cf. Granada, 2009, pp. 396–397.

[89] Vermij, 2007, p. 125, and 2002, p. 89. Near the end of his life, Lansberg published what may be considered "the first book in Europe which aimed at popularizing the Copernican theory among a non-mathematical audience." There, he put forward what Vermij describes as a series of occult and religious principles in support of the heliocentric hypothesis. See Vermij, 2007, pp. 122–123.

attributed different parts of the offspring to the heavens and the earth, they equally relied on the Holy Spirit as a spark to ignite their generative activity. Without it, their creative capacity lay "dead and buried" like an egg without the warmth of a brooding hen.[90] This was not the only time that Lansberg expressed his astronomical views through biological imagery. He also compared the diurnal rotation of the earth to the survival strategy of "a living being" that sought as much of "the Sun's life-giving light" as possible.[91] Additional examples of this imagery further suggest that Lansberg's adoption of the heliocentric hypothesis affected his view of the earth far more than his view of the heavens. With Kepler, however, we witness a change in the other direction, applying earthly knowledge to understanding celestial mutability.

Kepler accepted earthly knowledge as a starting point for celestial physics since he saw the two as essentially similar. Copernicus had "granted the right of citizenship in the heavens to the earth," Kepler proudly proclaimed, and there was no reason why the causes of optical properties "below the moon and among the fixed stars" could not be compared.[92] As Kepler wrote to his friend and rival Helisaeus Roeslin (1544–1616), he had "studied hard and shown" that there was "a much closer relation between the heavens and earth" than Aristotle had found.[93] In fact, the similarity of celestial objects with sublunar ones had encouraged Kepler to make physical claims about the heavens in the first place. If Roeslin wished to claim that the earth was different from the heavens "entirely in essence," Kepler advised his friend with a hint of humor to consider the greater ontological meaning of the earth's new place among the other planets:

> Dr. Roeslin further errs that the earth is distinct from the heavens entirely in essence and in location. I reply that Dr. Roeslin must first ask Copernicus where he calls home in the world. For the earth itself is in the heavens, moving in between Mars and Venus as if placed in the middle of the course, as Archimedes tells us that Aristarchus said 2,000 years ago.[94]

---

90 Ibid., p. 125. Cf. Matthew 23:37.

91 Vermij, 2007, p. 130.

92 *JKGW*, 1, 246.23–24.

93 *JKGW*, 4, 107.34–36. Roeslin adopted a geoheliocentric world system similar to that of Tycho. As Kepler summarized Roeslin's support of Tycho sarcastically, "Roeslin says that I split too many hairs and that nature is simple. Thus, the new foundations of Tycho are all the more agreeable, since they are meaner and more fitting of nature. Those of Copernicus, on the other hand, are more subtle and less fitting of nature." See ibid., 139.29–31.

94 Ibid., 109.1–5: "Ferners jrret Rößlinum / das die Erd vom Himmel tota essentia et loco gescheiden. Antwort. D. Rößlin muß erst Copernicum fragen wo er in der Welt daheimen sey. Dann die Erd selber ist im Himmel / und laufft mitten zwischen Marte und Venere daher / als gleich in die wet / in medio cursus constituta sagt der uralte Aristarchus vor

In the *New Astronomy*, Kepler had shown that the location of the earth, "moving in between Mars and Venus," was accompanied by an essential similarity with the other celestial bodies. Looking back on his book, Kepler pointed to "a magnetic virtue [*virtus magnetica*]" as the principal similarity among them.[95] The earth was not "entirely different in essence from the sun and the other planets," Kepler wrote, since the same magnetic virtue accounted for the orbit of the moon as well as the path of the earth and the other planets around the sun.[96] Here was persuasive evidence, Kepler claimed, which resolved Roeslin's objection to the essential continuity of the cosmos.

With this continuity in mind, Kepler deployed the bulwark of his celestial physics in the *New Astronomy* and *On the New Star*. Today, we consider the latter the "more quirky and much less 'modern'" of the two,[97] yet Kepler saw it more as a cosmological counterpart. *On the New Star* further conveyed his physical conception of the heavens, equipping the heliocentric hypothesis with a more complete cosmology. Many of the new features of this cosmology, notably the "physically motivated centrality" of the sun,[98] were part of a larger plan to bring Copernicus's discovery to fuller fruition. In pursuit of this plan, Kepler took the opportunity to resolve issues which had been raised against the new cosmology. One of these, the "immeasurable expanse" which Copernicus had suggested between Saturn and the sphere of the fixed stars, elicited a response whose "process of argumentation" spanned Kepler's two books.[99] In the *New Astronomy*, Kepler wrote on reflection, he had proven the annual

---

2000 Jahren / wie Archimedes das bezeuget." On Kepler's running debate with Roeslin, see Granada, 2005 and 2011b.

95 *JKGW*, 4, 109.8.

96 Ibid., 109.6–9: "As to whether the earth is essentially different from the heavens, it is not essentially different from the sun and the other planets entirely. Rather, the earth participates with them in a magnetic virtue [*virtutem magneticam*], especially with the moon. My book about Mars [that is, the *New Astronomy*] thoroughly confirms this."

97 Gingerich, 2002, p. 236. Gingerich also identifies *On the New Star* among those important works by Kepler that still await translation into English. In an earlier essay, Gingerich argues that "any complete assessment of Kepler's place in astronomy would necessarily notice his *De stella nova* (1606)." See Gingerich, 1975, p. 261.

98 Gingerich, 1993, p. 333.

99 *JKGW*, 4, 109.29–39: "The natural philosophers [*physici*] will now object and say that it is an incredible thing that the earth should revolve [around the sun]. And since they cannot understand our astronomical evidence, much less overturn it, they set on what follows from it and say, 'It is incredible and absurd that such an immeasurable expanse should exist between the moving and motionless stars. And since this is not the case, neither can that other thing be true, namely that the earth revolves [around the sun].' I faced these physical issues in my book on the new star of 1604, where I did not wish to admit that

orbit of the earth "through many visible signs of evidence."[100] Combined with the absence of stellar parallax, however, the earth's orbit implied the existence of "an immeasurable expanse between the planets and the fixed stars."[101] Kepler dealt in detail with this distance in *On the New Star*, where he continued his campaign for the causal superiority of the new cosmology.[102] At stake was the physical structure of the cosmos, along with the aesthetic principles of harmony, order, and proportion.[103] Not to be taken lightly, these principles were considered by Kepler as the marks of the master craftsman, whose divine image informed the human mind. The challenge for Kepler was to identify a "probable proportion" in this structure, which situated the sphere of the fixed stars "3,000 times higher than [the orb of] Saturn, and Saturn 12,000 times higher than the orb of the earth."[104] Such a proportion spoke to the essence of the human soul, Kepler claimed, and it confirmed our role as researchers who spoke the mathematical language of our creator. With this and other sources of evidence, Kepler gave full expression to his celestial physics in *On the New Star*, stressing our place of privilege as the students of creation.

Central to Kepler's celestial physics was an effort to reconfigure the cosmos according to what Miguel Ángel Granada calls "a horizontal relation."[105] By abandoning the hierarchical view of the active heavens and the passive earth, Kepler applied a single causal system to flatten, so to speak, the two spheres. Once conceived by Aristotle as "the source of all motion" for earthly events,[106] the heavens were now considered by Kepler comparatively, according to the same causes occurring on earth. We remember one result of this new relation when we celebrate Kepler for being "the first astrophysicist" responsible for "mechanizing and perfecting the world system."[107] We must not identify Kepler's system of celestial physics

---

it is such an incredible and absurd thing that the world should be as large as Copernicus wishes and I showed with examples that it is nothing new."

100 Ibid., 109.23.

101 Ibid., 109.26–27. Cf. ibid., 110.35–37: "There is no proportion too great for God that it should not come to pass. We men, however, do not perceive such proportions and are frightened for no reason by the immeasurable magnitude."

102 On the formulation of Kepler's 'Copernican campaign' in correspondence with his patron, Johann Georg Herwart von Hohenburg, see Boner, 2011a.

103 On Kepler's aesthetic appraisal of astronomical hypotheses, see Jardine, 2009, pp. 273–274.

104 *JKGW*, 4, 110.32–41.

105 Granada, 2011a.

106 Aristotle, 2004, p. 9.

107 Gingerich, 1993, pp. 305, 333.

exclusively with his account of planetary motion, however. If we credit him with being the first "to envision astronomy as a part of physics,"[108] we must also consider his broader view of physics involving a variety of vitalistic speculations. As we have seen in his study of the new star of 1604, Kepler suggested that superfluous material in every corner of the cosmos was the object of a faculty constantly at work in the conception of new and useful forms. Although he did not study this faculty with full certainty,[109] Kepler claimed that this "industrious architect" could bring about change in the heavens in the same way that it did on earth.[110] The two spheres were balanced out in this way by a similar cycle. Everywhere the process of corrosion led to a new form that contributed to the continual restoration of the cosmos. In the heavens, Kepler located the faculty responsible for this renewal in "the empty liquid [of the ether]," where it could generate comets and new stars in the same way that living beings on earth spontaneously produced new forms of life:

> For either there is no place for this faculty, except in the empty liquid [of the ether]; or, if it is in the globes of the stars, it will have many places; or, if it is in the globe of the sun, it will have an extremely small place and its projection [to the sphere of the fixed stars] will be extremely great, similar to infinity. I prefer to say the first, that there is throughout the ethereal substance one faculty where the planets follow their course and another faculty where the fixed stars remain in place, similar to the natural faculty that is in living beings.[111]

Attributing a generative faculty to the celestial ether, Kepler painted a world picture whose laws of planetary motion were set against a backdrop of sudden and unpredictable change. Although the spontaneity of this activity sharply contrasted with the regularity of his planetary apparatus, Kepler assumed that every novelty in the heavens stemmed from the same general spectrum of secondary causes. Produced by a physically causal process, the purpose of such a novelty was not the recovery of any particular thing, but the reversal of decay. Kepler took over the idea of

---

108 Stephenson, 1994, p. 2.

109 Before considering the efficient cause of the new star, Kepler began by saying, "I shall not contend much on such an uncertain matter." *JKGW*, 1, 267.10.

110 *JKGW*, 1, 268.37–38.

111 *JKGW*, 1, 269.9–14: "Aut enim sedes ei nulla, nisi in ipso liquido inani: aut si in globis stellarum; sedes ei non una: aut si in globo solis, sedes ei angustissima, excursus immensus, infinito similis. Primum dicere malo; inesse in tota substantia aetherea, unam, qua planetae decurrunt; alteram, qua fixae stant, facultatem similem naturali facultati, quae est in animalibus ..."

spontaneous generation as a common biological theory among his intellectual contemporaries. According to this theory, old and superfluous matter served as the start of an "instauration."[112] From fleas and ticks to frogs and fishes, an "industrious architect" lay behind the transformation of expiring forms into new ones:

> Whenever this faculty encounters any superfluous matter, it converts it into a living being that serves the nature of things, either by adding or taking away... The philosophers call this putrefaction, which is in fact nothing other than the instauration of old material from an expiring form, as if dying, to some new one, which is produced by this industrious architect, whoever he may be.[113]

In his study of the effects of this faculty in the heavens, Kepler relied heavily on Aristotle. Kepler recalled the "vital heat throughout the universe" which had suggested to Aristotle that all things were ensouled.[114] A similar sign of vitality could be sensed in "our warm air,"[115] which was widely considered an ingredient of fertility, together with moisture. On this, Kepler quoted directly from the *Generation of Animals*, where Aristotle had identified the elemental ingredients of species spontaneously produced from putrescent matter. Beginning with the elements of earth and water, Aristotle had attributed these forms of life to a sort of spiritual agency accompanied by a "vital heat."[116] In his study of animal reproduction, Aristotle had associated this spiritual agency with 'pneuma,' or the source of fertility in semen.[117] Aristotle had described the 'pneuma' as "more divine

---

112 *JKGW*, 1, 268.37.

113 *JKGW*, 1, 268.26–38: "... illa, quae quoties invenit superfluam aliquam materiam; convertit eam in animalculum tale, quod rerum naturae serviat, seu juvandae seu exonerandae... Philosophi putredinem dicunt, quae revera nihil aliud est, quàm materiae, pereunte et quasi moriente forma veteri, ad novam aliquam instauratio, facta per hunc operosissimum architectum, quicunque tandem is sit." Cf. ibid., 268.28–35: "... the sweat of the head into lice; the sweat of the body (especially of women) into fleas; the perspiration of every tree into a particular sort of caterpillar; the oily excesses of the skin, left over and losing their form, into moths; the oily excesses of marshland into eels; and those of the waters into frogs, fishes, leeches, and always an infinite variety of new sorts of sea monsters; the earth into various species of plants and trees. In sum, this is the source of flies and swarms of bees and beetles, hornets, and wasps and extraordinary flutters of butterflies."

114 *JKGW*, 1, 267.28–29.

115 *JKGW*, 1, 268.25–26. In his account of the material vehicle for this faculty, Kepler wrote that "it appears in either case to be a vapor which is born in the body and exhaled into the surrounding exterior, filling up the more hidden recesses of bodies with its moisture where heat is enclosed and allowed to endure."

116 Aristotle, 1942, pp. 356–357.

117 Ibid., pp. 170–171.

than the elements" and analogous to the ethereal substance of the sun.[118] Since the 'pneuma' acted as "a principle of life" in the same way that the heat of the sun effected generation,[119] the heat of living beings was seen as something more than merely fire. For Aristotle, the 'pneuma' and the sun similarly served as an efficient cause in the conception of animals and plants in every part of the earth. The comparison of these two sources, one in the heavens and the other on earth, conveyed a sense of continuity between the two spheres. For Kepler, one kind of world suggested one kind of knowledge.

Aristotle had recognized rotting fluids as a source of fleas and flies and putrescent soil as a source of plants, among a variety of other creatures that came into being "by some spontaneous activity of nature."[120] In the instance of animal reproduction, Aristotle regarded semen as "surplus residue," something like "the paint left over on an artist's palette."[121] To any other natural residue, Aristotle also attributed "a principle of life" that could participate in the process of production.[122] Proposing a similar principle in the heavens, Kepler suggested the fertility of residual material everywhere in the celestial ether. Just as Aristotle had identified "a principle of life" in the natural residue of animals,[123] so Kepler saw fertile grounds for new luminaries in the liquid substance of the ether. Kepler supposed the new star of 1604 had emerged from superfluous matter in the heavens in the same way, for example, that eels were born from bogs or worms from the oils of the skin. In this way, the new star was arguably the product of "oily and impure vapors" emitted by the other stars.[124] Kepler claimed the new star served as a sort of cleansing agent, preserving the transparency of the ether and contributing to cosmic renewal. From animal excess to vaporous areas of the celestial ether, Kepler found in spontaneous generation an analogous explanation for the new star.

118 Ibid.

119 Ibid., pp. 172–173.

120 Ibid., pp. 8–9.

121 Ibid., pp. 80–81.

122 Ibid., 170–173.

123 Ibid., pp. 172–173. In his summary of Kepler's stellar theory, Humboldt wrote that Kepler, "like Tycho Brahe, believed that the new stars [of 1572, 1600, and 1604] had been conglomerated from a cosmical vapor filling the regions of space." Curiously, Humboldt associated Kepler's cometary theory "with ancient fancies on spontaneous generation" without noting that his stellar theory shared a similar connection. In any case, Humboldt claimed that such considerations, "even when based on feeble analogies and not on actual observations, riveted the attention more powerfully . . . than the most important results of calculating astronomy." See Humboldt, 1997, pp. 327–329.

124 *JKGW*, 1, 269.30.

Spiritual Agency in Kepler's Astrology

Kepler's account of the new star brings us to his astrology, where he adopted another vitalistic view as the basis for his system of celestial influence. Rather than accept the transfer of motion as the main form of influence, Kepler suggested a sublunar soul responsible for discerning the disposition of the heavens and responding to their mathematically meaningful configurations. In the case of astrometeorology,[125] the role of the heavens in the weather was a matter of understanding how the soul of the earth responded to these configurations. Known as the aspects, these arrangements represented for Kepler a metaphysical form of continuity between the heavens and earth. As Kepler explained, when two or more planets formed an aspect, they exemplified a geometrical proportion that evoked a meteorological response. Essentially, every aspect was "a geometrical harmony" whose effect was felt "objectively rather than subjectively," since "the principle of motion" was in the earth and not in the heavens.[126] Kepler regarded every aspect as a relation of terms, "a thing of reason" sensed by the earth instinctually in the same way that a song moved someone who did not grasp the mathematical principles of music.[127] In a fanciful analogy, Kepler suggested that a similar thing would occur if "a sweet love song" were to move "an eager maiden" so strongly that she would become pregnant without any physical contact.[128] Although extraordinary, this analogy spoke to the reality of the earth's response to the appearance of the heavens. Stirred in the same way that music moved the ignorant listener, the earth responded to the realization of mathematical harmony in the heavens through heightened meteorological activity.

Kepler made clear that the harmony of the heavens ultimately belonged to the earth. That resonance was critical to making mathematical sense of the stars, as the earth compared the positions of the planets and perceived

---

125 In his description of astrology, Oresme defined astrometeorology as that part which dealt with "the revolutions of the stars and the conjunctions of the planets" as a way of predicting "the state of the atmosphere, the changes in the weather, from hot to cold, from dry to wet, winds, storms, and such movements in nature." See Coopland, 1952, p. 53. On the other areas of astrology, see the opening of Chapter Two.

126 *JKGW*, 4, 140.40–43: "An aspect is a geometrical harmony [*geometrica concinnitas*] between the light rays of two planets here on earth. A geometrical harmony is a relation, a thing of reason. A relation is not an effect, and so it follows that the aspect can produce no rain on its own."

127 Ibid., 141.3–4: "... nature is not moved in this way by an aspect subjectively but objectively, like a song moves the farmer to dance."

128 Ibid., 111.25–27.

their proportions.[129] The heavens, after all, had no harmonic principle in themselves. Rather, the soul of the earth was responsible for comparing their light rays and revealing their geometrical relations:

> For the earth alone knows and senses when there is an aspect between the light rays of two planets. The heavens, or the stars that are said to aspect with one another, know less than nothing of this, and as little as the organ pipe knows of the song which it makes possible.[130]

As this musical analogy also suggests, Kepler believed that astrology and music shared some of the same archetypal principles that had participated in the foundation of the world. The only difference between the two was that astrology was "radial," that is, it relied on the reflected light rays of the planets rather than the sounds of instruments and voices.[131] In view of this similarity, Kepler sought to reveal the rudiments of creation through the study of astrology along with other mathematical avenues. These avenues were fundamentally geometrical, Kepler argued, and found "especially in Euclid."[132] Geometry gave us "things of reason," Kepler wrote, and since reason was eternal, "the truth had been in the mind of God forever," like "an archetype of the world [*mundi archetypus*]."[133] From an early stage in his career, Kepler saw astrology as part of a larger harmony whose expression he would later articulate in the *Harmony of the World* (1619). As he explained in an early preview of this work, astrology figured as one of five parts of his project, along with arithmetic, geometry, musical theory, and astronomy. By uncovering the "causes of the aspects," Kepler claimed he would take a crucial step towards uncovering the archetypal beauty of

129 On the "necessary presence of the soul" in Kepler's later view of harmony, see Escobar, 2008, pp. 31–34. In his analysis of Kepler's theory of the soul, Escobar argues that Kepler considered it one of three elements necessary for knowledge, together with the body and the world. Although the terms of any harmony were found in the world, the realization of it required—and ultimately belonged to—the soul.

130 As Kepler wrote in response to Roeslin, "the cause of the aspects does not depend altogether on the reception of the earth nor on the influence of the heavens [*influxum coeli*] at all... The heavens are not excluded by me [from the efficacy of the aspects], however, since they provide the light rays for it. Dr. Roeslin should also grant me the earth, however." See *JKGW*, 4, 112.1–6: "Dann die [Erde] weißt und empfindet allein / wann es ein Aspect ist zwischen zweyer Planeten Leichtstraalen: hiervon waist der Himmel / das ist / die Sternen / quae se mutuò dicuntur aspicere, weniger dann nichts / und eben so wenig / als wenig die Orgelpfeiff von dem Liedlein weiß / darzu sie verhülffich sein muß."

131 *JKGW*, 15, no. 332, 64.

132 *JKGW*, 15, no. 357, 166–167. On Kepler's rejection of numerology, see Field, 1984a.

133 *JKGW*, 15, no. 357, 167–172.

the cosmos.[134] By 1606, he felt he had made significant progress on this front. This configuration defined Kepler's natural philosophy as well as his career.

As Kepler stated in a summary of *On the New Star*, he had already revised all of astrology according to "harmonic doctrine," removing from it everything that had obscured the underlying mathematical principles of celestial influence.[135] Ever since the completion of his *Cosmographical Mystery*, in fact, Kepler had rejected the twelve signs of the zodiac, the houses, and every other element that had stood in the way of the aspects. Kepler reported on his reform of astrology in a letter to his skeptical contemporary, Thomas Harriot (1560–1621):

> In 1596, I rejected the division [of the zodiac] into twelve equal parts, the houses, the rulerships, the triplicities, along with everything else, retaining only the aspects and relating astrology to harmonic doctrine. You will find my account in my little book, *On the New Star*, which is now being issued in Frankfurt.[136]

---

[134] *JKGW*, 14, no. 148, 11–19: "Although I am powerfully disturbed by the perilous state of affairs [in Styria], I have just now prepared the method and initial outline of a book, a cosmographical dissertation to which I shall give the title *On the Harmony of the World*. There shall be five books or chapters: 1. Geometrical, on the constructible figures; 2. Arithmetical, on the solid ratios; 3. Musical, on the causes of the harmonies; 4. Astrological, on the causes of the aspects; 5. Astronomical, on the causes of the periodic motions."

[135] *JKGW*, 15, no. 394, 76–77. In her survey of Kepler's astrology, Field explains that Kepler later abandoned any attempt to explain the aspects "by means of consonances." She notes, however, that Kepler would continue to regard the aspects as "natural 'harmonies,' though in a non-musical sense." See Field, 1984b, p. 207. On Kepler's theory of the aspects in Book 4 of the *Harmony of the World*, see ibid., 1988, 127–142; cf. Simon, 1979, 169–174.

[136] *JKGW*, 15, no. 394, 74–78: "Ego iam à decennio divisionem in 12 aequalia, domus, dominationes, triplicitates etc. Omnia reiicio, retentis solis aspectibus et traducta astrologia ad doctrinam harmonicam. Videbis meam sententiam ex opusculo, de stella nova, quod iam prodit Francofurti." Kepler later described in greater detail what he meant by "relating astrology to harmonic doctrine." See Bialas, 1983, p. 47: "Since four of the seven divisions of the circle corresponding to the seven harmonic divisions of a chord form the aspects commonly observed by astrologers, that is, the sexangular (sextile [60°]), the quadrangular (quadrant [90°]), the triangular (trine [120°]), and the diametral (opposition [180°]); and since experience continually shows that these figures are capable of inciting sublunar nature as often as the rays of two planets intercept an equivalent part of the zodiac, no less than the corresponding divisions of a chord are capable of sweetly soothing the ear, I arrived at the opinion that the creator either took the laws for ordering the aspects from the harmonic ratios or adjusted the human ears, the judges of harmony, to the celestial aspects. And so it followed that we ought to observe three other aspects in the heavens in addition to conjunction, that is, quintile [72°], biquintile [144°], and semisextile [30°], just as in music there are seven other harmonic tones in addition to unison."

As Kepler explained, the astrological division of the heavens had begun with the Chaldeans and was no less arbitrary than "the division of the week into seven days according to the seven planets."[137] While this form of division could provide a convenient reference for telling the time, there were many other references in astrology that afforded only "one fantasy after another."[138] As in the case of the zodiac, Kepler saw these divisions simply as a reference with no relation to the actual essence of the world. The only relation, in fact, Kepler found in the aspects, which he accepted for their realization of certain geometrical principles. While the aspects echoed the creation in this way, cultural devices such as the zodiac were seen as the relics of human history that revealed their artificial nature through experience. The height of the summer heat in August was not produced by the position of the sun in the sign of Leo, Kepler wrote, but rather the accumulation of heat in the body of the earth beginning in June. "Yet what do the heavenly signs have to do with the dense and dull body of the earth?" Kepler asked.[139] Kepler's astrology cut to the core of his cosmology, where he identified archetypal principles that lay at the foundation of every area of his thought. That Kepler compared astrology with music so frequently reflected their common foundation at the heart of his causal account of the cosmos.

Aware of opposition to the Copernican cosmology by other astrologers, Kepler argued that the aspects did not require the adoption of a geocentric viewpoint. Rather than remain motionless at the center of the cosmos, the earth continued to react to the heavens as it revolved annually around the sun, preserving its place as the "principle of celestial activity."[140] An important point for the heliocentric hypothesis, Kepler's emphasis on the agency of the earth also expressed a foundational principle for every form of celestial influence. From astrometeorology to nativities, Kepler saw astrology as the sphere of activity of an animate faculty in the soul of

137 *JKGW*, 4, 141.27–31.

138 Ibid., 141.37.

139 Ibid., 141.33–37: "Neither the sign of the lion nor the sun is responsible for the height of the heat in August, when the sun is in Leo . . . Rather, the reason is the earth, which has absorbed the heat of the sun in June and, due to its density, kept it until August, heating up all the more. Thus, with the accumulation of heat in the earth the celestial cause always comes first and the dense material (the body of the earth) always follows. Yet what do the heavenly signs have to do with the dense and dull body of the earth?"

140 *JKGW*, 4, 111.1–5: "Although it is true that my view [that the earth acts as the principle of celestial activity] acquires a more imposing appearance when one accepts that the earth also moves, my meteorology was not founded on the motion of the earth. On the contrary, one may accept that the earth stands still and it shall hold all the same."

the earth and every human being. All of these souls occupied bodies in continual motion, yet they stood similarly at the center of their own circle of influence. As a central reference point, every sublunar soul sensed the surrounding heavens through a faculty whose reactionary function Kepler considered instinctual. Equipped with this faculty, the earth and human beings discerned the aspects and danced to their tune in a different way. Kepler described this dance as the accommodation of "commotions" to the configuration of the heavens.[141]

Threads of Continuity in Kepler's Cosmology

Kepler often encouraged others to consider the physical causes of astrology more closely. Understanding the actual effect of the sun on the seasons did not require reference to the zodiac,[142] Kepler argued, nor was there any basis for the association of the planets with particular signs. In his prediction of the weather, Kepler aimed to removed these forms of reference and restore astrology to the causal and empirical criteria of the rest of physics [*physica*]. That Kepler accepted astrometeorology as a part of physics does not suggest, however, that he envisioned his system of celestial influence as mechanical in the modern sense. As we have seen, Kepler applied his clock analogy to the process of planetary motion, a single though significant part of his celestial picture. Taken as a whole, his celestial physics moved beyond the margins of mechanistic thinking.[143]

By turning to biology, Kepler displayed the full range of his celestial physics. For every phenomenon, Kepler suggested a form of agency, aspiring to "an astronomy without hypotheses" through a complete and causal explanation of the heavens.[144] That Kepler considered vitalistic faculties among other forms of agency did not imply a break from his account

141 *JKGW*, 15, no. 332, 137–141: "On this, I agree so closely with [Cornelius] Gemma that I suppose there to be no need for any 'spirit of the whole world.' Rather, it is enough if, in the natural faculty of the earth as well as in the natural faculties of all human beings, there is inscribed a sense of celestial things and a function of accommodating the commotions of their bodies accordingly." This passage is taken from a letter to Wolfgang Wilhelm von Neuburg (1578–1653); for a full translation, see Boner, 2011b.

142 *JKGW*, 4, 137.33–36: "Although the sun alters our seasons on earth according to the four qualities as it travels through the twelve signs [of the zodiac], this does not occur in the order in which the signs are arranged together by the astrologers."

143 See Barker, 2006, pp. 1–2: "Because he formulated the laws of planetary motion later accepted as correct by mechanical philosophers, Kepler is still sometimes read as a nascent mechanical philosopher with a deplorable weakness for mysticism."

144 *JKGW*, 1, 6.9.

of planetary motion according to "a single magnetic corporeal force."[145] Kepler's celestial physics drew on a variety of forms of causal explanation, including Aristotelian biology, which explained change through the agency of living beings. The fruits of his physical astronomy extended from the causes of planetary motion in the *New Astronomy* to those of celestial mutability in *On the New Star*.[146] Begun after the bulk of the *New Astronomy* was already complete, *On the New Star* built further on Kepler's application of earthly knowledge.

Kepler's singular focus on the aspects in astrology suggests a form of cosmic continuity even more profound than his explanation of celestial change. Although the application of earthly knowledge in astronomy "redefined the relation between disciplines" and reflected an original attempt "to achieve a unified view of nature,"[147] it also signaled for Kepler a superior level of causality. Beneath the ebb and flow of normal existence, Kepler identified a metaphysical foundation whose principles served to link the heavens and earth seamlessly. While the sheer distance of the heavens suggested a certain degree of speculation for his celestial physics,[148] Kepler firmly believed that every part of the world picture had been informed by these archetypal principles. Mathematical in nature, they spoke to a form of continuity that ran more deeply than his deployment of biology. In the following chapter, we turn to Kepler's early formulation of these principles and their role in his reform of astrology. As we shall see, they spoke to the same metaphysical thread that ran through the entire fabric of his thought. We will further explore how these principles relied for their recognition on a faculty in living beings. A vitalistic bridge between the heavens and earth, this faculty would prove essential to Kepler's mature conception of cosmic harmony.

---

145 *JKGW*, 5, no. 325, 60.

146 On the full range of physics in the view of Georg Liebler (1524–1600), who taught Kepler natural philosophy at the University of Tübingen, see Methuen, 1998, pp. 193–197. Westman argues that Liebler "almost certainly" presided over Kepler's student disputation in favor of the heliocentric hypothesis. See Westman, 2011, p. 317.

147 Granada, 2009, pp. 398–399.

148 On Kepler's use of optical knowledge to explore the physical features of the heavens, see Boner, 2009b.

CHAPTER TWO

KEPLER'S EARLY CAREER IN ASTROLOGY, 1594–1599

In 1594, Kepler was summoned to the Protestant seminary in Graz before completing his studies in theology at the University of Tübingen. There in the Styrian capital, Kepler was responsible for teaching mathematics and publishing astrological calendars. At the time, the mathematical curriculum was commonly divided into four parts, arithmetic, astronomy, geometry, and music, together known as the 'quadrivium.' Astrology, an essential part of astronomy, involved the observation of causes and effects widely considered as the "ordinances of God."[1] It has been correctly said that astrology, as taught in the quadrivium, "was part of a complete philosophical system of the world,"[2] and Kepler certainly saw it that way. The basic principles of astrology, however, were applied in different ways, depending on different areas of interest: medical astrology relied on celestial configurations for the administration of medicine and the diagnosis of patients; judicial astrology foretold the fates of individuals, groups, and nations; electional astrology sought to 'seize the moment,' determining the best occasions for holding important events; natal astrology predicted, sometimes retrospectively, the accomplishments and characteristics of individuals according to their natal chart, their personal map of the positions of the planets at their time of birth; horary astrology explored particular queries according to the 'birth' of the question and the natal chart of the questioner;[3] and astrometeorology considered the sympathies, often realized in the form of weather conditions, between the sublunar and celestial spheres.[4] Whether for the treatment of illness, the prediction of political uproar, or the forecast of major meteorological events, astrology in much of early modern Europe was thought to put

[1] Methuen, 1998, pp. 77–78.

[2] Field, 1984b, p. 190.

[3] Girolamo Cardano (1501–1576) claimed that this form of astrology was "accidental, against our law, and ... the cause of many evil things." Cardano studied several queries, however, covering every subject from the life and death of a ruler to the sex of a baby. In his study, he often urged his readers also to take into account factors external to astrology. See Cardano, 1966, pp. 553–560.

[4] For an introduction to the various areas of astrology and their shared study of celestial influence, see Campion, 2009; Tester, 1987; Bouché-Leclercq, 1979.

practitioners in touch with divine decrees. And for scholars such as Girolamo Cardano (1501–1576), astrology was "more delightful and divine than any other discipline."[5]

At first, Kepler saw his position in Graz as a temporary obligation to fulfill before returning to Tübingen to complete his studies in theology.[6] Soon, however, he discovered that his research in astronomy could reveal another form of learned reverence. Inspired by the discoveries that led to his first major publication, the *Cosmographical Mystery* (1596), Kepler described himself as a priest who studied "the book of nature," the work of the divine creator glorified by the astronomer. Conceiving this role as complementary to the study of Sacred Scripture, Kepler invested profound importance in the study of the stars. He expressed this view in a letter to Herwart, where he made clear his goal as the glory of the creator:

> Yet I am of the opinion that, since we astronomers are priests of the highest God with respect to the book of nature, we do not promote the praise of the intellect but above all behold the glory of the creator. He who is convinced of this does not easily bring to light anything other than what he himself believes, nor does he abruptly alter anything in [astronomical] hypotheses unless he hopes that from them the phenomena can be demonstrated with greater certainty.[7]

Kepler's sketch of the astronomer's role contains a scathing critique of those whom he saw as the clever combiners of "ancient hypotheses and the new ones of Copernicus."[8] Kepler argued that astronomers such as Tycho Brahe (1546–1601) and Nicolaus Raimarus Ursus (1551–1600) had wrongly viewed their vocation as "outshining those two great luminaries, Ptolemy and Copernicus."[9] Their suggestion of a 'compromise system,' Kepler wrote, was a vain attempt to surpass these authorities "in the glory of invention" rather than faithfully grasp and glorify the work of their creator.

---

5 Cardano, 1966, p. 728; cf. Grafton, 2000.

6 Kepler's summons to Graz has been described as "a bitter personal struggle" that left the aspiring theologian "torn between his calling and his duty." See Voelkel, 1999a, p. 23.

7 *JKGW*, 13, 193.14–19: "Ego verò sic censeo, cum astronomi, sacerdotes dei altissimi ex parte libri naturae simus: decere non ingenii laudem, sed creatoris praecipuè gloriam spectare. Qui hoc sibi persuasum habet, is non facile aliud quicquam in lucem emittit, quàm secum ipse credit, nec temerè quid in hypothesibus mutat, nisi certius ex illis phaenomena demonstari posse speret."

8 Ibid.

9 Ibid. Kepler also named Philip Lansberg, Giovanni Antonio Magini (1555–1617), and Helisaeus Roeslin.

How did Kepler's divine vision of astronomy relate to his view of astrology? First, it is important to note that Kepler occasionally took great pains to differentiate between astrology and astronomy. At one point, Kepler described the popular practice of astrology as "a foolish daughter" whose "mother, the most reasonable astronomy," relied on predictive principles rather than the particular circumstances of people.[10] Those who pretended to project "their foreknowledge onto the future contingencies of individuals," Kepler proclaimed, were nothing more than "godless philosophers and physicians."[11] In 1599, Kepler briefly outlined what would eventually become the five books of his *Harmony of the World* (1619), corresponding to geometry, arithmetic, music, astrology, and astronomy, respectively. Describing the final, astronomical book as the study of "the causes of the periodic motions," Kepler assigned to astrology, the focus of the fourth book, the consideration of "the causes of the planetary configurations."[12] Second, despite the certainty Kepler so often attributed to mathematical knowledge, he appears to have allowed for plenty of uncertainty in astrology.[13] It may well be argued that Kepler's astronomy, burdened by a potentially boundless number of archetypal principles, was similarly plagued by the specter of incertitude,[14] but there can be no mistaking what Kepler claimed to be two very different degrees of demonstrability. Kepler often compared astrology to the imprecise art of the physician, advanced by the accumulation of knowledge about physical qualities. Just as the physician would continue gaining greater familiarity with "the many and various herbs," Kepler wrote, so would the astrologer continue developing "a discrete science [*scientia*] and body of knowledge" about the natural properties of the planets.[15] In either case, greater experience gave way to a measure of consistency that served as a basis for "bearing out the causes."[16] Whether or not Kepler held out the possibility of improving the precision of astrology, however, he did not anticipate

---

10 *JKGW*, 4, 161.9–17.

11 Ibid., 181.26–27.

12 *JKGW*, 14, no. 148, 18–19.

13 On the element of uncertainty in Kepler's astrology, see Rabin, 1987, p. 157.

14 On the role of archetypal principles in "Kepler's construction and defence of his physical astronomy," see Martens, 2000, pp. 97–98, 146–168. On the ambiguous identity of Kepler's archetypes, see Donahue, 2002, p. 297.

15 *JKGW*, 4, 177.18–25.

16 Ibid., 164.9–15.

perfecting his practice anytime soon. And the greater goal of arriving at the causes of influence remained even farther in the future.[17]

Yet despite the many differences Kepler identified between astrology and astronomy, he claimed the two shared in geometry the same metaphysical foundations. As a consequence, Kepler applied geometrical principles to the two areas by way of analogy.[18] He also extended these principles to the study of music. In fact, Kepler argued that all material phenomena, from the motions of the planets to the effects of the heavens on the weather to the production of particular melodies, derived from the same singular set of geometrical principles. Seen in this way, astrology, astronomy, and music shared the same archetypal origins. Kepler even described astrology as "a silent music" whose appreciation was made possible by a soul that could "dance to the tune of the aspects."[19] As Kepler made clear, the universal nature of geometrical principles accounted for the underlying consistency of the cosmos, where the idea of harmony encompassed far more meaning than our modern understanding. On the occasion of accepting three new aspects as influential, Kepler wrote to Herwart in 1599 on his discovery of an "absolute analogy" between astrology, astronomy, and musical theory:

> ... The analogy [*analogia*] with music and astronomy is absolute. I show that the analogy must necessarily be seen in this way, since the origins of all things are derived from geometry. Nature confirms these principles in the creation of a single species and employs these principles in everything that is capable of them. This occurs in music, the motions of the planets, the operation of the planets [on earth], the measure of musical notes according to time, the dances of men, and the composition of songs. For although these things are the discoveries of men, nevertheless man is the image of the creator.[20]

---

[17] Kepler also claimed he might not fully comprehend the causes of influence when he finally came to consider them. See ibid., 164.14–15.

[18] Originally intended by Plato to refer to a mathematical proportion (ἀναλογία), the word *analogia* came to signify in ancient and early modern Latin "a method of reasoning from parallel cases." See the *Oxford Latin Dictionary*, 1985, p. 126.

[19] *JKGW*, 11,2, 48.23–28; cf. Bialas, 2004, p. 141.

[20] *JKGW*, 14, no. 130, 640–651: "... perfecta sit analogia musices et astronomiae. Quam analogiam necessariò spectandam hoc medio demonstro, quia omnium rerum origines ex geometria petitae sunt, et quas natura rationes probat in creatione unius generis rerum easdem adhibet in omnibus omninò rebus, quae earum sunt capaces. Propterea in musica, in motibus planetarum, in operatione planetarum, in dimensione notarum musicalium causâ temporis, in hominum saltationibus, in ratione carminum. Nam etsi sunt haec hominum inventa, tamen homo imago conditoris est."

Kepler's association of astrology and astronomy was not only ontological. He claimed that humanity had been intellectually equipped with the very principles underlying both of these mathematical branches. And although these principles appeared to be human discoveries, their conception was no mere coincidence. Man, created in "the image of the creator," had found them as a consequence of his natural character. Kepler argued that knowledge of archetypal principles had not come about by choice, since "the measure of nature" was not determined by "the election of man."[21] Rather, archetypal knowledge was a reflection of our very essence. Seen in this way, the material world exemplified certain principles already present to our intellect. Stemming from the same general system, astrology and astronomy suggested a set of intellectual tools whose practitioners stood on the same epistemological scaffolding. It is for this reason that Gérard Simon would not ultimately distinguish between Kepler "the reformer in astrology" and Kepler "the renovator of astronomy."[22]

Not All Astrologers Created Equally: Kepler's Perception of His Practice

It is clear from his early correspondence that Kepler did not claim for astrology the accurate determination of anything.[23] He issued calendars annually and wrote at length on astrology, yet Kepler rarely passed up the opportunity to point out the many problems of his practice. Particularly problematic were those whom Kepler saw as the astrological counterparts of empirics and quacks in the world of medicine. Surrounded by so many promoters of falsehood, Kepler openly preferred to pursue "other, more agreeable and more vital studies" than astrology, despite his professional obligations.[24] Calendars and the composition of birth charts brought him material benefits, but he did not deem it "disgraceful to disregard certain axioms of the astrologers" which would doubtfully improve his own practice.[25] Kepler pointed out, for example, that previous scholars such as Johannes Schöner (1477–1547), whose astronomical tables featured a famous defense of astrology by Melanchthon, had chosen from among "an

21 *JKGW*, 13, no. 123, 366–367.
22 Simon, 1975, p. 447; cf. Kusukawa, 1995, p. 188.
23 On the general nature of Kepler's astrology, see Rabin, 1997, p. 753; 1987, pp. 191–192.
24 *JKGW*, 13, no. 123, 611–612.
25 Ibid., 612–614.

enormous magnitude of astrological volumes" when selecting their own "frivolous foundations."[26] Kepler considered much of astrology to have been corrupted by the contributions of countless practitioners whose efforts had obscured an original core of truth.

With this conviction in mind, Kepler compared his role in astrology to that of the religious reformers of the day. Like those reformers, Kepler sought to separate the chaff of human corruption from the kernel of truth. How he went about doing this, he explained, did not recall the methods of the Jesuits, whose missionary efforts had been mixed up with many frivolous practices. Rather, Kepler modeled his reform of astrology on the Lutherans, who had removed several trivial traditions in arriving at firmer theological footing. In a letter of 1598, Kepler made use of this comparison to gain the sympathy of his mentor and former astronomy instructor at Tübingen, Michael Mästlin (1550–1631). In an earlier letter, Mästlin had questioned Kepler's choice of context when expressing his critical views of astrology in his calendars. Quick to note the nobility of his cause, Kepler drew a parallel between his critique of astrology and the integrity of his Lutheran brethren:

> Noble instructor, can it be that I do not act properly if I devote my work to persuading scholars and even philosophers of a distinct operation of the heavens? Here, I act as the Jesuits, who amend many things in order to make men Catholics. Nevertheless, I do not act in this way, for those who defend every frivolity resemble the Jesuits. I am a Lutheran astrologer, who casts aside frivolities and preserves the kernel.[27]

Kepler's self-portrayal as a "Lutheran astrologer" may appear inopportune to us. Although Luther had allowed that some celestial configurations could herald the judgments of God, he had condemned astrology as incompatible with divine omnipotence.[28] It seems more appropriate to interpret Kepler's comparison according to the manner in which Luther's reform of the Christian religion related to Kepler's self-proclaimed role in the restoration of astrology. Kepler saw his simplification of astrology to a select set of geometrical principles—a return to astrology's roots—

---

[26] Ibid., 607–610.

[27] *JKGW*, 13, no. 89, 173–178: "Optime praeceptor, an non rectè ago, si operam do, ut doctis et philosophis etiam operationem caeli persuadeam distinctam? Ago itaque ut Jesuitae: qui multa emendant ut homines catholicos faciant. Imo non ita ago, nam qui omnes nugas defendunt sunt Jesuitis similes, Ego sum Lutheranus astrologus, qui nugis abjectis retineo nucleum."

[28] Dixon, 1999, pp. 408–409.

as analogous to Luther's removal of the many "frivolities" that Catholicism had contrived to conceal the original core of Christianity. Comparing the customs of Catholicism to the conventional rites of astrology, Kepler likened the idolatrous worship of saints to such things as the system of houses. In both cases, he claimed, human invention had obscured the original source of conviction. Whether or not we should consider Kepler's efforts as an attempt "to make astrology a natural science,"[29] it is clear that astrology shared with his other areas of interest an underlying objective to eliminate what he saw as strictly cultural conventions. Such things as the divisions, names, and various properties of the zodiac were identified as the astrological equivalents of what Kepler considered in Christianity the adverse accumulation of human custom. Kepler, like Luther, sought to strip away these customs, revealing a simplified body of sound doctrine.

Comparing his endeavors in astrology to the Lutheran reform of religion, Kepler frequently failed to convince others. His attempts were more often seen as outright assaults that would isolate astrology from its original foundations. As we have recently seen, Kepler was also prone to choosing the wrong places to express his critical views. In his letter of 1598, Mästlin had informed Kepler that the venue his former pupil had chosen to communicate his criticism was questionable. Mästlin and "not some few others" had wondered why Kepler had attacked astrology so strongly in his calendar of 1598, having made so little criticism in the *Cosmographical Mystery*, where Kepler "would have been able to argue only with the learned."[30] The "useless tools of the numbered houses and the arrogant presumption of predicting particular things" were expressly condemned in his calendar,[31] while Kepler had simply assumed these same things in select chapters of the *Mystery*. The result, Mästlin had claimed, was a calendar whose critical content fell on the deaf ears of his audience. Without objecting to Kepler's views, Mästlin had opposed the place his former pupil had chosen to air his opinion:

> For [the calendar] is done for country folk and the uneducated, who neither expect nor observe such arrogant presumption, but rather predictions about the annual harvest, war, and other things of this sort. In fact, those people

[29] Simon, 1975, pp. 440, 444. Cf. Field, 1984b, p. 220: "As we have seen, what seemed 'nonsense' and what 'kernel' did change over the years, so that the *Confessio Augustana* of Kepler's alleged astrological Lutheranism was more a collection of principles than a definite body of dogma."

[30] *JKGW*, 13, no. 97, 47–53.

[31] Ibid., 50–53.

> do not grasp the foundations of the principles of the twelve houses of the zodiac, the triplicities, etc. Yet I shall now keep quiet about the matter. You see that I have not opposed your opinion, only that it seems to me to have been put forward in an inappropriate place, that is, in a prognostic [*prognosticum*], which lasts a single year, while other written works, in which the material could have been discussed with greater facility, are everlasting.[32]

In his response to Mästlin later the same year, Kepler described the circumstances differently. With only 400 to 600 calendars published, Kepler saw little concern for his commentary reaching a wide audience. As a consequence, Kepler composed his calendars mainly for "nobles and prelates," who claimed for themselves "some knowledge of things" which they did not actually know.[33] The end result might still have been the same, that is, Kepler might still have not won any followers, but it is worth noting that he was willing to direct his efforts of reform to an audience of potential patrons. Kepler carefully distanced himself in his calendars from those astrologers whom he considered "frivolous," the consumers of convention, and focused on the few principles which he believed were based on reality rather than ritual.[34] If his approach won him few followers, it proved even less profitable when it came to producing patrons. As Kepler explained to Mästlin, his salary had seen no significant increase since he had begun composing calendars:

> In all of my prognostics [*prognostica*], I try to provide from ready convictions that seem to me true a certain taste of the delightfulness and greatness of nature to my readers ... If perhaps they are moved to support me with as great a salary, this has yet to come to pass ... If you accept these things, you shall no longer be angry with me, for I am a protector of astrology through my words and work, and yet I continue to strive to implant the opinion in the minds of men that I am not a frivolous astrologer.[35]

---

32 Ibid., 56–61: "Hîc enim cum rustica plebe et idiotis, à quibus non arrogans illa praesumptio, sed prognostica de proventibus anni, de bellis, et quae eiusmodi sunt, expectatur, aut spectatur, agitur. Etenim his ignota sunt fundamenta rationum 12 domorum coeli, triplicitatum, etc. Sed de his etiam quiesco. Vides me mentem tuam non impugnasse, sed quod mihi videatur, eam impertinenti loco fuisse propositam, nimirum saltem in prognostico, quae durant unum annum, cum alia scripta, quibus illa materia commodius tractari potuisset, sint perpetua."

33 *JKGW*, 13, no. 106, 61–63. For an analytical summary of Kepler's response to Mästlin, see Sutter, 1975, pp. 297–298.

34 Steven Vanden Broecke suggests that many proposals for astrological reform served as "boundary markers, through which elite practitioners distinguished themselves from competing 'popular' practitioners." See Vanden Broecke, 2003, pp. 262–270.

35 Ibid., 68–75: "In omnibus prognosticis id ago, ut de promptis sententiis, quae mihi verae videntur, gustum aliquem jucunditatis et majestatis naturae praebeam ... meis lectoribus ... si forte per hoc excitentur ad me tanto majori cum salario alendum, cujus nulla

Kepler did not explain to Mästlin why he chose to critique astrology in his calendar. Instead, he attempted to convince his mentor that he had chosen an appropriate context. Kepler did not believe in such "frivolities" as predicting political revolutions: the sequence of seasons was "one thing, humankind another."[36] Kepler felt it necessary to explain why he had spoken out against these things in his annual prognostic. Lamentably, his lack of frivolity in astrology was accompanied by an equally sparing financial state, and he appears to have held out little hope that his patrons, allured by the "delightfulness and greatness of nature" found in his calendars, would improve his circumstances notably. Yet despite the many things in his calendar that required an apology, Kepler refused to repress his opinions. His skeptical view of astrology stemmed from the same foundational principles at the heart of his cosmology. If many of his readers could neither understand nor appreciate the philosophical roots of his astrology, Kepler did not allow their interests to determine the depth or direction of his analysis.

To lend measure to the magnitude of his critical attitude towards astrology, Kepler expressed little interest in examining the growing literature on the subject. In a letter of 1599, Kepler compared his ignorance in astrology to the ignorance of Ursus in astronomy. Suspecting he was guilty of the same crime that Ursus had committed by mixing up the major hypotheses in astronomy, Kepler threw up his hands and simply surrendered to the possibility that he was unaware of widely held principles. "When I weigh the enormous number of volumes of the astrologers," Kepler wrote, "I suspect that I am doing the same thing in astrology that Ursus does in astronomy."[37] Through his comparison, Kepler accused Ursus of not actually reading those authors whose theories he had wrongly related in his book, *On Astronomical Hypotheses* (1597). Kepler claimed he was equally unprepared when it came to "some of the axioms of the astrologers."[38] The price of his ignorance, however, was far less than the damage done by Ursus. Kepler was occupied by other areas of study he described as "more delightful and less dispensable," in particular astronomy.[39] Kepler excused his ignorance in astrology as something he did not view

adhuc accessio facta ... Haec si probaveris, jam porrò mihi non succensebis, quod assertor astrologiae cùm sim verbis et opere, studeam tamen hanc animis hominum existimationem implantare, non esse me nugacem astrologum."

[36] Ibid., 78–80.

[37] *JKGW*, 13, no. 123, 607–611.

[38] Ibid., 613.

[39] Ibid., 612.

as disgraceful. If this was a matter of priority, he would not sacrifice his study of astronomy to cover the last couple of centuries of astrological literature. On the contrary, Kepler claimed that this literature only shrouded astrology further beneath the shadows of convention. Slimmed down and simplified, Kepler's astrology called for a return to the timeless principles of geometry apart from human history.

Through his close involvement in the controversy between Tycho and Ursus, Kepler would become closely familiar with Ursus's skeptical view of astronomy. Contracted to work for Tycho under the obligation that he provide a defense for his patron, Kepler would set out in 1600 to undermine Ursus's misrepresentation of astronomical hypotheses. Angered by Ursus's view of their validity, Kepler declared in his *Defense of Tycho against Ursus* that "false hypotheses, which together yield the truth once by chance, do not in the course of a demonstration in which they have been combined with many others retain this habit of yielding the truth."[40] When Kepler moved on to the observational equivalence of competing systems of prediction, he determined their differences according to physical considerations. "Thoughtless" astronomers were condemned for adhering strictly to the numbers and suggesting that "the same result follows from different hypotheses."[41] For Kepler, emphasis on the physical implications of astronomical hypotheses allowed him to distinguish between competing systems and curtail the criticism of skeptical adversaries such as Ursus. In his attention to qualitative considerations, Kepler attributed particular importance to the physical principles of coherence, simplicity, and precise archetypal patterns.

Tying Ursus's ignorance to his skepticism, Kepler condemned him of a crime whose consequences were far greater in astronomy than in astrology. Kepler's higher regard for astronomy is often the focus of historical fiction, where he is described as neglecting his astrological duties at the price of his annual pay.[42] We may even imagine the young Kepler struggling at times with his calendars while savoring his time spent on the "more delightful" astronomy. Although some historians have painted a similar picture,[43] it is important to note that Kepler questioned the con-

[40] Jardine, 1984, pp. 89, 140.

[41] Ibid., pp. 90, 141.

[42] See, for example, Luminet, 2009, p. 295, where Kepler is issued an official warning and "a fine of two florins for every day" his calendar of 1596 is delayed. Under the obligation of authority, Kepler is torn from his *Cosmographical Mystery* "like an angel hurled from the highest star to the mire of a meaningless zodiac."

[43] See, for example, Tondorf, 1904, p. 303.

temporary practice and not the validity of astrology. He allowed for a certain degree of ignorance of astrology because he doubted the propriety of popular practice rather than the reality of celestial influence. Kepler also saw little point in forecasting the actions of individuals whose fickle paths proved far less foreseeable than the perennial motions of the planets. As testimony to astrology, Kepler pointed to the influential properties of a particular group of geometrical archetypes. According to Kepler, the cosmos had embodied these archetypes since the original act of creation, allowing for a form of interaction between the celestial and sublunar spheres. Kepler claimed that these archetypes were exemplified by certain configurations of the planets, which, as a source of stimulus in the sublunar sphere, led to different responses in human behavior, disease, and other earthly events. How the earth and her human inhabitants responded to these configurations was a highly complex matter.

Conserving the Kernel: Kepler's Early Conception of the Astrological Aspects

What did Kepler mean by the metaphysical archetypes he saw as so essential to understanding astrological influence? We know that Kepler's view of these principles changed with time,[44] so how did he conceive of them after completing the *Cosmographical Mystery*? In April 1599, Kepler gave an answer to this question in a long letter to Herwart. Described as a "scholar of great energy, if little judgment,"[45] Herwart wrote regularly to Kepler through a network of correspondence involving several other potential readers. The letters exchanged by Kepler and Herwart were first dispatched to the imperial court in Prague, where various dignitaries could read them. An ardent Catholic and outspoken statesman, Herwart ensured that his correspondence with Kepler passed before the eyes of numerous non-Protestants. This arrangement has been seen as having "raised Kepler from the mass of his colleagues," earning him "special consideration with the ruling Catholic party."[46]

In a letter of March 1599, Herwart asked Kepler to help him sift through "the Arabic frivolities" surrounding astrology and find "the philosophical

[44] On Kepler's conception of the aspects over the course of his career, see Bialas, 2004, pp. 139–144.

[45] Grafton, 1991, pp. 187–189; 1997, p. 198.

[46] Caspar, 1993, pp. 69–70.

pearls."[47] Having recently come across a book by Heinrich Rantzau (1526–1598) and several other titles on the subject, Herwart solicited the services of the young mathematician as he made sense of the many authors who had written about astrology. Herwart also asked about their observational sources. In his response a month later, Kepler began by affirming the interaction of the celestial and sublunar spheres through ordinary observation. As he described it in a marginal note he made in Herwart's letter, this interaction involved a "connection between the heavens and earth and the effects flowing from it."[48] Kepler also mentioned the work of Aristotle as relevant to the matter, while Ptolemy, "led astray by the fables of fools," was described as "having neglected nature."[49] "Aristotle considers those philosophical pearls in his books *On Generation and Corruption* and *Meteorology*," Kepler wrote, "and I have pointed out some of them in the preface of my prognostic for this year, which you have."[50] Kepler also cited such things as the association of terrestrial vapors "with the light of the moon" and the sequence of the tides "according to the motion of the sun and the moon" as evidence of an influence from beyond the lunar orb.[51] As his most significant source of evidence, however, Kepler suggested the aspects, those configurations of the planets whose effects on the sublunar sphere derived directly from the metaphysical archetypes.

In his letter to Herwart, Kepler outlined eight influential configurations of the planets. Also known as 'aspects' or 'planetary radiations,' these configurations conveyed influences "by way of reason."[52] Seen as the center of a circle whose circumference corresponded to the zodiac, the earth served as the principal reference point for the planets. Although Kepler accepted the motion of the earth, it remained always at the center of this circle while the planets continued moving around the circumference. Latitudinal motion did not play a part, as Kepler conceived the circle, and consequently the positions of the earth and the other planets, in terms of two dimensions. When two planets became separated by a certain distance or angular section with respect to the central reference point, an aspect was momentarily formed. Since the planets proceeded on their paths almost continually, an aspect often passed into obscurity as soon as it came into

47 *JKGW*, 13, no. 114, 110–113.
48 Ibid., 221–222.
49 Ibid., 220–225.
50 *JKGW*, 13, no. 117, 193–196.
51 Ibid., 201–203.
52 Ibid., 203–204.

existence. Kepler listed the aspects in order of ascending angular separation: conjunction (0°), sextile (60°), quintile (72°), quadrature (90°), trigon (120°), sesquiquadrature (135°), biquintile (144°), and opposition (180°).[53] Each of these aspects found a musical counterpart among "the eight harmonies."[54]

Five of the eight aspects were originally outlined by Ptolemy,[55] the leading authority of early modern astrology.[56] These configurations, conjunction, sextile, quadrature, trigon, and opposition, were the cardinal configurations. According to Kepler, they surpassed in influence the three new aspects he had introduced, quintile, sesquiquadrature, and biquintile.[57] These new aspects were based on "the same reasoning" that underlay "those time-honored and traditional ones."[58] Kepler called on natural philosophers to inquire after the validity of every aspect by observing the weather regularly. The sudden onset of certain conditions at the sign of an aspect, he claimed, was clear evidence that the heavens played a part in earthly affairs. Kepler had collected evidence in favor of the new aspects since 1594, having compared their occurrence "with recorded storms."[59] This influence did not result from the light radiated by the sun or reflected by the planets, however, but from the nature of their separation at a particular point:

> You see, today when two planets are separated by 89°, nothing new occurs in the weather [*meteoris*]. Tomorrow, when they are separated by fully 90°, that is, [when they are] in quadrature, a storm suddenly arises. What little light arrives from these two planets in a single day, and how can it be reduced so suddenly tomorrow? Thus, this is not the effect of any number of stars nor of light, but the sum of 90°, that is, the angle from the rational number 90°, a harmonic part of a complete circle.[60]

---

53 Ibid., 209–211.

54 Ibid., 206–207.

55 See Ptolemy, 1940, pp. 72–73.

56 On the importance of Ptolemy in Melanchthon's transformation of the educational curriculum in Lutheran institutions of learning, see Methuen, 1998, pp. 97, 219.

57 *JKGW*, 13, no. 117, 211.

58 Ibid., 212–213.

59 *JKGW*, 14, no. 132, 351–352.

60 *JKGW*, 13, no. 117, 213–219: "Ecce hodie cum distant planetae duo 89 gradibus, nihil novi fit in meteoris. Cras cum distant plenis 90°, sc: quadrante, subito oritur tempestas. Quantula luci utriusque facta est accessio intra unum diem, et quomodo perendie illa minui statim rursum potest? Igitur non stellae est is effectus, sed stellarum, non lucis, sed numeri 90°, hoc est anguli per numerum 90° rationalem, et harmonicam totius circuli partem ..." The term *meteora* has been taken to refer to 'the conditions of the weather.' Here, it appears to stem from the original Greek '*μετέωρα*,' 'rising up, levitating,' which referred

Kepler argued that an angle produced by the separation of two planets by "the sum of 90°" led to a change in atmospheric events. The metaphysical nature of this phenomenon suggested to him "a soul in the earth fit for understanding those aspects [*apta ad intelligendos hos aspectus*]."[61] Although Kepler employed observational support for the three new configurations, he considered this metaphysical line of reasoning as more compelling evidence.

In a later letter to Herwart, Kepler linked each aspect to one of the eight ratios that were held as harmonic, unison (1:1), minor third (6:5), major third (5:4), fourth (4:3), fifth (3:2), minor sixth (8:5), major sixth (5:3), and octave (2:1).[62] Kepler made this connection by comparing the measure of the circumference cut off by a particular configuration with the circumference of the circle as a whole. The larger part of this proportion measured the whole of the circumference, while the smaller part measured the segment separated by the circumferential span of two planets (including the sun and the moon). Sextile, for example, cut off 60° or $^5/_6$ of the circumference, corresponding to the harmonic ratio of 6:5, the minor third. Opposition, which cut off 180° or $^1/_2$ of the circumference, resulted in the harmonic ratio of 2:1, the octave. The three new aspects, quintile (72°), sesquiquadrature (135°), and biquintile (144°), rounded out the group of eight and referred to the ratios of 5:4, 8:5, and 5:3, respectively. Kepler also illustrated in his letter to Herwart how the aspects achieved this agreement. Straightening the zodiacal circumferences of the eight aspects into a line, Kepler equated the angle of every planetary configuration with the ratio formed by a particular harmonic proportion.

In a letter of May 1599, Herwart questioned Kepler's account of the aspects. Diplomatic in tone, Herwart expressed doubt about rejecting "everything in astrology except for the planetary aspects."[63] Since "even the smallest foundation" in astrology would seem "to depend on sheer

---

to every phenomenon regarded by Aristotle as an atmospheric occurrence produced by a mixture of moist and dry exhalations. In addition to those phenomena we continue to consider atmospheric, Aristotle classified comets, meteor showers, and shooting stars as *meteora* (Aristotle, 2004, 340b, 341b, 359b–360b, 370a). Although Kepler condemned this view of comets as "doubtful and untrue," he accepted all of the other *meteora* originally identified by Aristotle. On the multiple meanings of *meteora* in ancient Greece and Rome, see Taub, 2003, pp. 1–2.

[61] *JKGW*, 13, no. 117, 222–224.

[62] On Kepler's early coupling of the aspects and the consonances and his eventual rejection of their correspondence, see Simon, 1979, pp. 44–46.

[63] *JKGW*, 13, no. 121, 41–43.

superstition,"[64] Herwart urged Kepler to explain why he held on to what appeared to be only another false part of astrology. Herwart could simply not see why "many, if not an infinite number of aspects" could be introduced according to the same principles as Kepler's original eight.[65] Failing to find any logical way of limiting the number of aspects, Herwart invited Kepler to explain their distinct nature in greater detail. "I shall now keep quiet," Herwart finally wrote, "and leave out those things that can be raised against that refuge of writing horoscopes."[66] Despite his tact, Herwart did little to hide his doubt. Kepler would elaborate on the matter in the same place he would explain the common foundation of the aspects and the musical consonances.

Admittedly, the prospect of an infinite number of aspects had previously occurred to Kepler. He claimed to have resolved the matter in the *Cosmographical Mystery*, where he had distinguished the aspects from the twelve signs of the zodiac, the artifacts of "human design" that had "no basis in nature."[67] There, Kepler had also shown how the aspects held "something in common with the consonances."[68] Yet despite his reference to the *Mystery*, Kepler felt compelled to explain to Herwart why he accepted only eight aspects. "You object," he wrote to Herwart, "that there are an infinite number of aspects just as there are nearly an infinite number of harmonies."[69] In his letter of May 1599, Kepler acknowledged the possibility of an endless number of aspects as a problem previously discussed by several prominent authorities, including Ptolemy. At stake was the very distinction of the aspects according to their natural causes:

> If five aspects operate along with an infinite number, then the doctrine of the aspects is worthless. For Ptolemy even gave a cause for why men should have chosen those chiefly and omitted the others, but he did not prove that nature herself does not possess any more. The choice of man is not the measure of nature.[70]

---

64 Ibid., 47–51.

65 Ibid., 45–47.

66 Ibid., 47–48.

67 *JKGW*, 1, 39.22–23.

68 Ibid., 43.24–26.

69 *JKGW*, 13, no. 123, 372–373.

70 Ibid., 363–367: "Si 5 aspectus operantur, infiniti operabuntur, Vana est igitur doctrina aspectuum. Nam Ptolemaeus causam quidem dedit, cur homines hos potissimum elegerint, caeteros omiserint, at non probavit, naturam ipsam plures non habere. At non est Naturae mensura hominis Electio."

Once again, Kepler argued that the essence of the aspects was to be found in nature rather than the literature that had so often led us astray. As for the practice of prediction, Kepler recommended that Herwart remain general. "I do not move beyond the prediction of physiognomy, temperament, and major medical crises," Kepler wrote, "and in this way I consider myself free from superstition."[71]

Why did Kepler think the number of aspects had remained unresolved for so long? While he recognized the existence of no more than eight harmonic divisions of a string,[72] it was not their correspondence with these divisions that determined the eight aspects as influential. More significantly, the aspects shared with the consonances the same metaphysical foundations, a core set of archetypal principles constituted by clearly defined constraints. Kepler claimed these principles had produced a set of regular polygons that could be inscribed in a circle with the help of a ruler and compass. In the case of quadrature, the separation of two planets by 90° on the circumference of a circle formed the side of a square whose vertices touched the interior of the circle at four equidistant points. With trigon, the angular displacement of 120° amounted to the side of an inscribed triangle whose three vertices split the circle into three equal parts. The numbers of angles and sides of these figures were not as important to Kepler as their ability to be constructed rationally. According to this central premise, polygons such as the heptagon were prevented from being accepted.

Kepler considered the inscription of regular polygons as a way of accessing archetypal principles. These principles were geometrical patterns made momentarily manifest by the disposition of the heavens. Kepler would later define the regular polygons as figures whose "sides and outward-facing angles" were equal and whose inscription in a circle was determined by "the proportion of the side of the figure to the diameter of the circle."[73] "To know in geometry," Kepler later wrote, "is to measure by a known measure, which for the purpose of inscribing figures in a circle is the diameter of the circle."[74] For Kepler, the knowability and constructability of these figures were two sides of the same coin. And since irrational figures could not be constructed rationally, Kepler could not accept their influence in the form of celestial configurations. The eight aspects

[71] Ibid., 430–433.
[72] Ibid., 374–375.
[73] *JKGW*, 6, 20.17–21.25.
[74] Ibid., 21.30–33.

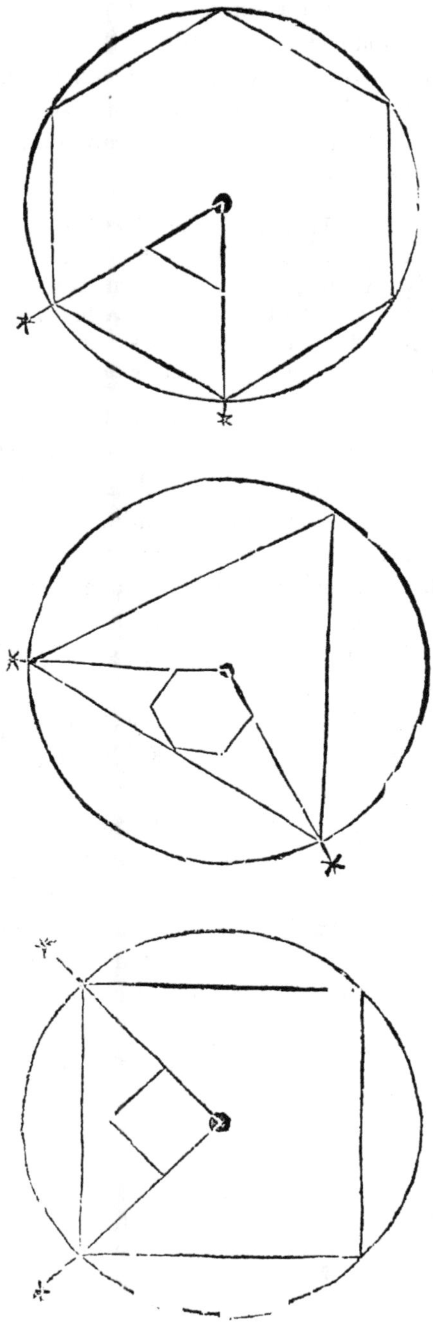

Figure 1. Agreement of quadrature (90°), trine (120°), and sextile (60°) with the sides of a square, triangle, and hexagon respectively. Johannes Kepler, *Harmonices mundi liber IV. De configurationibus harmonicis radiorum sideralium in Terra* (Linz, 1619), pp. 145–146. Courtesy of the Dibner Library of the History of Science and Technology, Smithsonian Library Institutions.

he did accept were based on the same archetypal principles that underlay the harmonic consonances. In his letter to Herwart, Kepler compared the geometrical grounds of the aspects to a dwelling place whose three new residents still required scrutiny before gaining admission to the inner abode of the original five. The infinite other configurations, meanwhile, were kept completely outside:

> Thus, either a cause must be produced for why the five aspects should operate and not the other three, or the other three must be admitted into the company of the five. Although this may happen, however, the door through which the other three are admitted lies in a dwelling place where the house remains closed even when the door to the site is open. And the crowd of countless aspects is kept outside, shut out by the front door.[75]

Among the countless mass of aspects "shut out by the front door," Kepler accepted only a few according to a select set of geometrical principles. That the doctrine of the aspects was "worthless" was thus proven wrong by the use of these principles. The limits of rationality, represented by our inability to realize certain figures, prevented the vast majority of configurations from ever being considered for candidacy. Kepler had followed a similar line of reasoning with his polyhedral hypothesis in the *Cosmographical Mystery*, where the number of regular polyhedra set a natural limit on the number of planets.

Kepler further explored the archetypal roots of the aspects in another letter to Herwart of August 1599. There, Kepler suggested an absolute analogy between astrology and music that Herwart could grasp if he considered their common cause more closely. Variously realized in the form of regular polygonal principles, geometry underlay astrology, music, and every other area of interest for Kepler. In the area of astrometeorology, Kepler attributed these principles to a sublunar soul, "an operative force [*vis*] of the stars" on earth that accounted for the influence of the heavens on the weather.[76] For Kepler, the study of astrology summoned the same set of principles that he found at the heart of his epistemology. Unified by these archetypes, Kepler's cosmos was based on a divine metaphysical

---

[75] *JKGW*, 13, no. 123, 377–380: "Quare, aut causa confingenda est, cur 5 aspectus operentur, 3 non: aut 3 residui in consortium admittendi sunt. Id etsi fiat, tamen janua quâ admittuntur, intra aedes est: quâ patefactâ, domus nihilominus adhuc clausa est: populusque infinitorum aspectuum foris stat, januâ summotus exteriori."

[76] *JKGW*, 14, no. 130, 579–581.

blueprint.[77] That man was created in the image of the creator, he claimed, allowed us to understand this blueprint together with the formal causes of natural occurrences.

Despite Kepler's best efforts to convince Herwart of the aspects, Herwart remained doubtful. As he confessed in a letter of August 1599, he still could not see the aspects as anything more than the products of opinion and superstition. While Herwart could accept other features of the cosmos as the consequence of "some particular and special reason," he saw no hint of substance in the aspects in judicial astrology:

> I do not doubt that it happened for some particular and special reason that among all of the fixed stars only seven should be found moving. And yet I would still not see any firm foundation from which that force and effect commonly attributed to the aspects of the stars should be proven, such that also all of astrology (judicial, that is) seems to me to depend entirely on opinions and superstitions.[78]

In the face of something that seemed so slippery and uncertain, Herwart chose to reject the aspects along with the rest of astrology. He simply could not understand how such a dubious subject had attracted so many while far greater questions had gone without significant study. Why had the creator chosen to create the heavens in such a way, he asked, with "so many stars in the firmament of such and such a size, separate from one another" in such and such a way?[79] And why had God chosen to create "seven moving stars and not fewer," following their paths in such a way?[80] "I marvel that there has never been any place or people," Herwart wrote, "who either did not discuss these two particular questions among celestial matters or did not convey them more fully to posterity."[81] Without offering any further explanation, Herwart affirmed the complete rejection of astrology. Although he recognized that "some numbers of the consonances coincide remarkably well with the numbers of angles that

---

[77] Rabin, 2010, p. 62: "This was Kepler's view of the natural world: ordered, comprehensible, reducible to a single principle, as the divine creator's world ought to be."

[78] *JKGW*, 14, no. 133, 60–65: "Nec dubito. Ex speciali, et praecipuâ quadam ratione id evenisse, ut inter tot fixas, non nisi septem errantes reperiantur. Tametsi uerò nondum adhuc ullum firmum fundamentum uiderim, ex quo vis illa, et effectus, qui aspectibus stellarum tribui solet, comprobari possit, uti quoque tota Astrologia (Juditiaria videlicet) ex meris opinionibus et suspitionibus mihi quidem dependere videtur."

[79] Ibid., 69–70.

[80] Ibid., 70–72.

[81] Ibid., 65–68.

subtend the sides of figures inscribed in a circle,"[82] he did not make any further commitment. And if Herwart did not accept the aspects, Kepler stood little chance of convincing him of anything at all.

From the Earth to Humanity: Further Effects of the Astrological Aspects

Kepler emphasized not only the effects of the aspects on the weather. In his letters to Herwart, he also stressed their significance for the constitution of human characteristics. Moving from astrometeorology to judicial and natal astrology, Kepler argued that the configuration of the heavens at the time of birth left a lasting influence, suggesting a celestial stamp that led to particularities in an individual's personality. According to this original imprint, everyone acquired a certain set of behavioral tendencies. In addition to serving as a source of temperament, this configuration became a sort of scale that could measure the impact of future configurations. Comparing the nativity to gardening devices used for shaping gourds, Kepler stressed the ability of the aspects to shape—rather than determine—the daily activities of individuals:

> How does the configuration of the heavens at the moment of birth become the character of a man? For it operates in him as long as he lives, not unlike those restraints bound round gourds by the skillfulness of farmers. Although they do not give life to the gourd, they do shape it. In the same way, although the heavens may not give a man his morals or actions or fortune or children or wealth or a wife, they do shape everything befalling him.[83]

Although the heavens would pass into a series of "infinite forms" over the course of life,[84] Kepler focused on their original impression. The study of a nativity was a question of "something that is not there,"[85] an exploration of the enduring influence of the heavens on one's soul. The body was "far too coarse to take up this character," Kepler claimed, while the soul was well suited for serving as the subject of it.[86]

---

[82] Ibid., 57–59.

[83] *JKGW*, 13, no. 117, 224–230: "Qua ratione facies caeli in puncto nativitatis, fit character hominis. Operatur enim in hominem quamdiu is vivit, non secus ac compedes illae injectae cucurbitis, agricolarum ingenio: quae cum cucurbitam non vegetent, tamen formant. Sic caelum etsi nec mores nec facta nec fortunam nec natos nec divitas nec uxorem homini det, omnia tamen homini obvenientia format."

[84] Ibid., 230–231.

[85] Ibid., 231–232.

[86] Ibid., 309–313.

Kepler saw this original imprint as a personal reference point that conditioned the actions and events in every individual's life. The key part of this personal watermark was the aspects. Kepler found a familiar example of their influence in the life of Barbara, his wife, whom he had married in April 1597. Jupiter and Venus had been poorly positioned at Barbara's birth, and their lasting influence had left her plump and plain-minded:

> You may observe an individual in whose nativity those noble ones, Jupiter and Venus, are unfavorably positioned ... You may see that such an individual, however honest and sensible, enjoys a rather unfortunate and sorrowful fate. I know such a woman well. She is praised by the entire city for her virtue, humility, and modesty, while she is equally simple-minded and corpulent in frame.[87]

Commended by the community for her moral qualities, Barbara bore flaws that her husband associated with the unfavorable influence of the heavens. Kepler went on to assign a similar influence to the series of sad marriages that Barbara had suffered. Born to a prosperous family, Barbara had already endured two miserable marriages before meeting the "poor and contemptible" Kepler. As he confessed to Herwart, Kepler only worsened her condition by earning a small salary and achieving little success in securing a share of her inheritance:

> From an early age, she was treated harshly by her parents and, scarcely an adolescent, married a forty-year-old man beyond the pale of pleasure. When he died, she immediately married another man of the same age with a more active spirit, though he was not much of a man and spent the full four years of their marriage suffering from illness. The third man she married is poor and contemptible, she [having been] wealthy before.[88]

Kepler moved on from marriage to the many other adversities encountered by Barbara. Plagued by the original configuration of Jupiter and Venus, she gave birth "with difficulty" and struggled even with simple activities.[89] Tying these other details together with her simple-mindedness and series of meager marriages, Kepler claimed that Barbara bore "the same

---

[87] Ibid., 236–241: "Videas hominem, in cuius genesi non commodè siti sunt boni illi, Jupiter et Venus ... talem igitur hominem videas, quamvis probum et sapientem, invenustiori tamen et subtristi ut plurimùm fortunâ uti. Talis mihi nota faemina est. Laudatur tota urbe ob virtutem pudorem modestiam. Simplex tamen juxtâ est, et crasso corpore."

[88] Ibid., 242–247: "Haec ab ineunte aetate duriter habita a parentibus, vix adolescens nupsit quadragenario praeter lubitum, eo statim mortuo nupsit alii eiusdem aetatis alacriori animo, sed qui neque vir fuit, et totum quadriennium, quod in hoc vixit matrimonio, per morbos exegit, tertio nupsit pauperi et contempto, dives ipsa antea."

[89] Ibid., 248–249.

character of body, soul, and fortune, clearly an analogy with the configuration of the heavens."[90]

Kepler accounted for his own character according to a similar analogy. At his birth, Saturn and the sun were separated by 60° in the angular subtension of sextile. From his fervent interest in philosophy to his feeble frame, Kepler linked everything to this configuration, down to his very diet. It was at the heart of his sardonic sense of humor and his penchant for "bitter and pungent things," suggesting a strong saturnine element in his life:

> My customs are similar [in character to my body and soul]. I gnaw on bones and cram down dry bread. I enjoy bitter and sour things and walking through hills, ruts, and thickets. I neither possess nor desire any charms in life aside from my studies, and I refuse gifts.[91]

Kepler considered all of the above customs in concert with the location of Saturn at his birth. He even speculated that his disposition, "drawn against society," drove him to challenge the human race by arguing in favor of the motion of the earth.[92] Setting in motion "a globe of such great weight, gliding swiftly through the stars,"[93] was certainly a sign of audacity, and the original imprint of the heavens might have pushed him farther down the path of rebellion. In this way, Kepler identified with other luminaries such as Cicero, whose resistance had also provoked the general populace.

By offering these personal examples, Kepler attempted to provide proof of how an original configuration could forever affect the actions and circumstances of an individual. Considering the coincidence of certain events with the appearance of the aspects at birth, Kepler argued it was "impossible that the soul be the forger of fortune entirely."[94] Rather, the heavens would also leave their mark on one's life events by painting them in a particular hue. In this way, an original configuration would color every decision and event in the course of life, setting the body, soul, and fortune to the tune of the stars. If the body was "the image of the world," Kepler wrote, then the origin of every individual found a counterpart in their

---

[90] Ibid., 249–250.
[91] Ibid., 257–260: "Mores consimiles. Ossa rodere, siccum panem ingerere, amara, acerba gustare mihi deliciae, per salebras per clivos per dumeta ambulare festivitas. Delinimenta vitae, praeter literas, nulla nec habeo nec desidero, et oblata respuo."
[92] Ibid., 264–268.
[93] Ibid., 266–268.
[94] Ibid., 250–252.

ascendant and a demise in their descendant.[95] "This shapes the actions of an individual," Kepler explained, "as well as those things that occur while the individual is inactive."[96] From the body to the soul and even to external circumstances, the principal elements of the human condition were in part the product of the heavens.

Kepler also told Herwart how the original impression of the heavens conditioned an individual's response to subsequent configurations. Every arrangement after birth, he claimed, could be better understood by comparing it with this original mark. Assuming that "a natal configuration" would endure in every soul, Kepler related it to "the real stars revolving in the heavens" as a way of making sense of their role in one's life.[97] Accounting for the past, present, and future, this comparison revealed the basic contours of every biography from beginning to end. Returning to his own life as an example, Kepler noted the happy coincidence of one particular configuration at a point of medical crisis, "a nearly fatal fever" that had healed with the help of the heavens at the age of sixteen. As Kepler explained to Herwart, this configuration had been foretold by the form of the heavens sixty-four minutes after his birth:

> One hour after the moment of birth, the appearance of the heavens has aged by fifteen years, with 15° elapsing in an hour. An hour later, their appearance has aged by thirty years. In this way, the first six hours may extend to ninety years of one's life. At intermediate points, similarities are found among one's actions and other life events and those things rising and setting in the sky. In my own case, when a nearly fatal fever had befallen me at the age of sixteen, nothing more convenient could have happened than what was signified by Mercury in opposition and Mars in quadrature in 8° Cancer, that is, an hour and 4′ after my ascendant in rising above the horizon.[98]

Here, Kepler described how the heavens had appeared similarly sixteen years after what they had shown just over an hour after his birth. He marveled at the relation of the original imprint of the heavens and their

95 Ibid., 296.

96 Ibid., 299–301.

97 Ibid., 389–392.

98 Ibid., 347–355: "Hora una post momentum partus, lapsa est in caelo imago 15 annorum, sicut in caelo 15 gradus caelestes horâ una moventur, horis 2, imago 30 annorum, adeo ut horae sex primae ad 90 hominis annos se porrigant. Quaeque intermediis horarum momentis in caelo oriuntur vel occidunt, eorum similitudines in actionibus aliisque temporaneis reperiuntur. Mihi anno 16 cum febris penè lethalis obvenisset, convenientius fieri nihil potuit, quàm ut ea significaretur ab ☍ ☿, □ ♂ in 8 ♋, hora una et sc: 4′ post meum ortum exoriente supra horizontem."

influence after birth, wondering how their original appearance could endure with "the changing sky."[99]

Still, Kepler could think of more spectacular things among the stars than the enduring influence of their original appearance. Managing to endure even more years, their motion was all the more remarkable:

> We marveled earlier that the configuration and form of the heavens may endure anywhere in the changing sky. Now, let us marvel all the more that the motion of the heavens is able to endure anywhere, allowing it to operate after so many years. For we understand their configuration and form to be imprinted in the same way as a seal is imprinted on wax, although that is an actual thing.[100]

Applying a classic metaphor, Kepler compared the impression of the heavens to the mark of a signature ring on sealing wax. This metaphor did not quite capture reality, however, since the soul was thought more subtle than any material phenomenon. Kepler searched for another example in vain. The image of something could "stay in the eye" long after the object had disappeared, but Kepler believed these images were "an actual thing."[101] Turning from sight to sound, Kepler noted that noise was "the image, as it were, of struck bodies," and the motion of the heavens did not produce any. "We have never seen nor heard an example," Kepler finally admitted, even though the matter had still not been "examined enough."[102]

Despite his best efforts to convince Herwart of the influence of the heavens on human lives, Kepler was unsuccessful. The kernel of truth that Kepler attempted to retrieve relied on the acceptance of certain principles that lay at the core of his cosmology. Herwart failed to grasp the rational limits determined by the construction of these principles. As Kepler later explained in the *Harmony of the World*, they set a natural limit that disqualified any figure whose side, in proportion to the diameter of a circumscribing circle, could not be determined "geometrically" from the number of angles of that figure.[103] By not accepting the geometrical rules that Kepler laid down as limiting principles, Herwart allowed for an infinite number of archetypal possibilities. As a result, he saw no reason

---

[99] Ibid., 369–370.

[100] Ibid., 369–372: "... mirabamur antea situm et faciem caeli permanere alicubi caelo abeunte. Iam multo magis est, ut miremur motum caeli alicubi manere posse, ut post tot annos operetur. Situm enim et faciem alicubi imprimi intelligimus, ut in cera sigillum imprimitur quae res est."

[101] Ibid., 375–376.

[102] Ibid., 374–380.

[103] *JKGW*, 6, 22.28–31.

to limit the number of aspects to eight or any other number, for that matter. Herwart was equally apprehensive about the influence of the aspects, which he viewed with as much uncertainty as every other part of astrology. Failing to appreciate the role of the archetypes in Kepler's cosmology, Herwart could not understand why the aspects would remain the one essential element of Kepler's astrology. Rather than dwell on the aspects, however, Herwart preferred to move on to more pressing matters relating to the motives of the creator.

In his response to Herwart, Kepler argued that any question about the cosmos inevitably involved the same set of underlying principles. This held true whether Herwart wished to examine the number and displacement of the stars or how the heavens acted on earth. (On the first question, Kepler claimed to have given a perfect explanation for the number of planets in the *Cosmographical Mystery*, where he had removed "any doubt" there were six in motion around the sun.)[104] In a letter of December 1599, Kepler illustrated the comprehensive nature of his cosmology in an outline of a "cosmographical dissertation" that he would later publish as the *Harmony of the World*.[105] As described above, Kepler foresaw five "books or chapters" corresponding to the areas of geometry, arithmetic, music, astrology, and astronomy, respectively.[106] With subject matter as diverse as "the causes of the harmonies," "the causes of the aspects," and "the causes of the periodic motions [of the planets]," those books would be brought together by the same geometrical principles.[107] Through their unity, Kepler proposed a form of disciplinary harmony that differed sharply from growing political and religious divisions in his province. Strongly disturbed by the rising tide of the Counter Reformation in Styria, Kepler barely managed to complete an outline of this work before abandoning Graz. Drawn to Prague by the promise of a position at the new observatory there, Kepler left Graz in 1600.

The Weight of Proof: Observational Evidence for the Astrological Aspects

Kepler did not rely exclusively on archetypal principles for his account of the aspects. To determine their influence on the weather, he assembled

104 *JKGW*, 14, no. 134, 510–512.
105 *JKGW*, 14, no. 148, 13.
106 Ibid.
107 Ibid., 15–19.

a collection of observations from his early career. In turn, these observations allowed him to measure the accuracy of his annual prognostics.[108] In a letter of 1595, for example, Kepler confirmed his forecast of an unusually harsh winter for that year. As evidence of "an extraordinary cold," he recounted the experience of local herdsmen who had recently returned home only to find their noses frozen and falling to pieces.[109] "So far, my calendar is correct," Kepler wrote, "for many herdsmen in the Alps perish from the cold when they return home and wipe their noses."[110] Nor was there any doubting the influence of the heavens in the spring of 1599, when an unusual drought and blazing sun had burned through the month of April. "Is it really chance," Kepler asked, "that there was a long-lasting quadrature of Saturn and Jupiter?"[111] Emboldened by this evidence, Kepler often emphasized the importance of "astrological experience" as a sure way of distinguishing "arbitrary causes" from the demonstrable influence of the aspects.[112] He even asked Herwart to accept the testimony of the weather as a primary way of fostering faith in their influence. In the light of observation, "the doctrine of the aspects" was simply "not to be rejected."[113]

Kepler thus combined the supposition of certain archetypal principles with the support of observational evidence. He claimed a configuration that did not possess the right geometrical properties could not be confirmed by experience. And once the right principles were in place, he invited others to probe their potency empirically. Kepler argued that the principles he had found called for the addition of three new aspects whose influence was worth scrutiny. In effect, Kepler had put forward a binding principle, a new and unique basis for why the eight aspects were "the only ones."[114] While he praised Ptolemy for filtering many "frivolities" and focusing more on nature, Kepler scorned the Alexandrian astronomer for arguing that the original five aspects were "the only ones taken into consideration by us."[115] Claiming reason rather than custom, Kepler rejected the many practices he thought had "led astray astrologers and

---

108 For an introduction to Kepler's annual prognostics in his early career, see Boner, 2008a.

109 *JKGW*, 13, no. 16, 16–20.

110 Ibid., 16–19.

111 *JKGW*, 14, no. 134, 498.

112 *JKGW*, 4, 138.7–141.39.

113 *JKGW*, 14, no. 134, 491.

114 *JKGW*, 13, no. 117, 422–425.

115 Ibid., 412–424.

natural philosophers" from the firm footing of the aspects.[116] His was a complete return to nature, a path that revealed three new configurations and a revolutionary connection with the world of music. In later correspondence, Kepler would continue to rely on his observations in defense of these discoveries.[117] Let us now turn to a later example of Kepler's use of "astrological experience" together with the archetypal principles.

In March 1608, Kepler's account of the aspects would come under attack from the physician and friend Johann Georg Brengger. Intrigued by Kepler's mention of "meteorological experience," Brengger asked Kepler for specific examples in support of the three new aspects. Given the number and variety of configurations that came at such a brisk pace, Brengger could not associate certain weather conditions with the appearance of the heavens at any particular point. The result, Brengger claimed, was a confusing flood of configurations that called into question the very possibility of proving the aspects:

> You write that you have proven the extraordinary aspects [quintile, biquintile, and sesquiquadrature] from meteorological experience. I should like to see an example of this experience, for with so many different aspects occurring so often that you may not be certain which one you attribute a change in the weather, I do not know how to make use of experience or if it is even possible.[118]

Brengger did not believe he could derive any evidence for the aspects from observing the weather. In the end, he challenged not only the application of experience but the very ability to distinguish a discrete number of influential configurations.[119]

In his reply to Brengger, Kepler turned to the relation of the aspects and the musical consonances. Building on his earlier battle with Herwart, Kepler explored the nature and number of the aspects according to the

---

116 Ibid., 419–422.

117 On the importance of observational evidence in Kepler's various accounts of the aspects, see Field, 1988, pp. 61–63, 128–130.

118 *JKGW*, 16, no. 480, 6–10: "Scribis te aspectus extraordinarios et ab experientia meteorologica comprobatos habere, optarim ego huius experientiae specimen videre, nam in tanto aspectuum numero et varietate, qui semper occurrunt, ubi incertus sis cui eorum mutationem aliquam aëris adscribas, nescio quomodo experimentum capere debeam, vel etiam possim."

119 Brengger became a regular contact of Kepler in December 1604, when Brengger suggested that they accept "philosophy as a means of establishing a friendship." Impressed by Brengger's articulate and careful criticism in correspondence, Kepler asked him in October 1607 to read the manuscript of his forthcoming *New Astronomy*. See *JKGW*, 15, no. 310, 2–6, and *JKGW*, 16, no. 448, 1–7; cf. Caspar, 1993, pp. 171–172, 266.

same basic principles at the heart of his theory of harmony. "You see that the consonances in music and the aspects in astrology," Kepler wrote, "spring from the same geometrical source of the division of the circle."[120] And even though the consonances were found on a chord and in this way "more removed" from their circular source,[121] they shared the same characteristics that kept the aspects from extending to infinity. Kepler gave his weather observations priority of place, however, beginning his letter to Brengger with an example of the influence of quintile (72°), one of the new aspects. Naming just a single example, Kepler noted there were many more. He also made it clear that his example had been calculated according to the ephemerides of Giovanni Antonio Magini (1555–1617) and confirmed with the observational instruments of Tycho:

> I may give very many examples of experience of the secondary aspects, but there is not enough space for describing my observations. For in the year 1600 when none of the primary aspects appeared from 23 April to 2 May (N.S.) and quintile was displayed by Saturn and Jupiter according to Magini, an abundance of snow fell in Prague on 1 May and in Styria during the wedding ceremony of [Archduke] Ferdinand... Through direct study of the heavens, it was found that on those same days there were 72° between Saturn and Jupiter. For Tycho's students provided proof of this for me with a Tychonic quadrant.[122]

Kepler suggested an even greater significance for observational evidence in considering a new configuration. Semisextile, the circumferential separation of 30°, seemed opposed to "all reason," yet Kepler was willing to accept it on "experience alone."[123] Here, the observations indicated a source of influence in the absence of reason. Along with the prospect of another aspect, Kepler faced the risk of reconceptualizing the relation of astrology and music. To complicate matters further, the same criterion of experience would call into question sesquiquadrature, one of the other new aspects already accepted on archetypal grounds:

---

[120] *JKGW*, 16, no. 488, 42–44.

[121] Ibid., 44–46.

[122] Ibid., 8–15: "Specimina experientiae de aspectibus secundariis, plurima dare possem, sed non vacat describere observationes meas. Anno quidem 1600. cum à 23 Apr. in 2 Maii St: No: nullus esset ex primariis aspectibus, ⚓ uerò ♄ ♃ in Magino exhiberetur, 1 Maii copiosissima nix cecidit et Pragae et in Styria in nuptiis Ferdinandi... Consulto coelo, inventum est iisdem diebus inter ♄ ♃ esse 72° gr: Nam studiosi Tychonis in meam gratiam periculum fecerunt quadrante Tychonico." Archduke Ferdinand married Maria Anna of Bavaria, daughter of Duke Wilhelm V, on 23 April 1600.

[123] Ibid., 93–95.

> I confess that reason supplied me with those three, quintile, biquintile, and sesquiquadrature, while experience alone supplied semisextile against all reason … Experience confirms quintile, biquintile, and semisextile, while it confirms sesquiquadrature more obscurely and doubtfully.[124]

It is testimony to the importance of observational evidence that Kepler was moved "against all reason" to consider semisextile as an influential configuration. Whether he could ultimately limit the number of aspects to eight, he could not deny their tangible effects on earth.

As for Brengger's reluctance to match the weather to the aspects, Kepler argued that it was not particular changes so much as "a general commotion of nature" that the stars could stimulate.[125] Predictions based on the arrangement of the heavens at a particular point could only be made in a general manner due to regional differences and other details that relied on a local knowledge of nature. These factors did not reduce the value of astrology, Kepler claimed, but reinforced it as the study of the regional resonances of the heavens. Returning to his example of heavy snow in the spring of 1600, Kepler explained that the same configuration could have produced another condition. The sign of influence, he suggested, was the sudden uproar of snow rather than the particular state of weather that he could not have easily foreseen:

> Yet I attribute nothing of note to the aspects except for a general commotion of nature that generates the weather. For what in this example was snow can in another be rain and similar things: it can also be a dry wind, it can be a wet vapor, which, with the faint appearance of a cloud in the bright light of the sun during the day, is the forerunner of fissures.[126]

Whether in rain, wind, or snow, Kepler identified the influence of the heavens in the weather. Setting aside any specific association, he focused on a fundamental resonance in the form of meteorological phenomena. This was a symphony orchestrated by the aspects, which produced a palpable, if often unpredictable, effect on earth. In the following chapter, we explore this symphony further as it unfolds in Kepler's response to a brilliant new luminary in the heavens.

---

124 Ibid., 93–125: "… fateor, tres hos ⚓ ★ # ratio mihi suppeditavit … sed semisextum nudissima experientia contra omnem rationem … Experientia confirmat ⚓ ★ ✳ sed # confirmat obscurius et dubie."

125 Ibid., 15–17.

126 Ibid., 15–20: "Sed nota nihil aspectibus tribuo, nisi in genere commotionem Naturae quae gignit meteora. Quod enim in exemplo nix fuit, in alio potest esse pluvia et similia: potest et ventus esse siccus, potest humida esse exhalatio, qualis de die resplendet instar nebulae, humilis et luce Solis clara, praenuncia chasmatum."

CHAPTER THREE

THE NEW STAR OF 1604

Forced by regulations of faith to leave Graz in September 1600, Kepler departed the Styrian capital with an uncertain future. Having written anxiously to Mästlin in the hope of help, Kepler received only words of advice from his mentor with an apology and prayer that "God preserve, protect, and watch over" his family "and all of the faithful."[1] Shortly before leaving Graz, Kepler wrote one last letter to Mästlin outlining his plans to join Tycho in Prague. Promised a position at the new observatory underway there, Kepler had run out of salary and left Graz in a hurry. He made his way hastily to Linz, where he would return if his position in Prague did not pan out. (From Linz, Kepler would travel with his family to Tübingen, where he hoped Mästlin might still help him.) Meanwhile, Kepler feared the swift dissipation of his wife's estate, in danger of falling into the hands of Catholic officials if he did not sell it within forty-five days of the decree of Archduke Ferdinand (1578–1637). "The entire substance of her estate is in property," Kepler wrote, "which carries so little value that it is not even sellable."[2] Caught in the middle of so much uncertainty, Kepler saw merit in his material sacrifice. Beyond the suffering of his family and "some few brothers," he believed spiritual compensation would come according to the scale of his sacrifice:

> Yet I would not have thought it so sweet to suffer in the company of some few brothers injury and indignity for religion, for the honor of Christ, to abandon house, home, field and friend. For if it is the case with true martyrdom and the sacrifice of life that the exultation is in some way greater, the greater the loss, then it is also easy to die for religion.[3]

Once in Prague, Kepler sought out Tycho, who had been appointed Imperial Mathematician in 1599. Although Kepler had traveled to Prague without confirming his place there, Tycho promptly secured him a salary at

[1] *JKGW*, 14, no. 153, 36–37.

[2] *JKGW*, 14, no. 175, 48–50.

[3] Ibid., 52–56: "At non credidissem, adeò dulce esse, pro religione, pro Christi honore, cum aliquantulo caetu fratrum, damna, contumelias pati, domos, agros, amicos, patriam deserere. Si verum martyrium et vitae jactura, proportione quadam respondet, ut quo maius damnum hoc maior laetitia sit: facile est et mori pro religione." Cf. Caspar, 1993, p. 115.

court. Along with the primary task of preparing a new set of astronomical tables named after the Holy Roman Emperor Rudolph II, Kepler was asked to write in defense of Tycho in his legal dispute with Ursus.[4] Kepler soon took up these tasks, tackling the orbit of Mars under the supervision of Christian Longomontanus (1562–1647) and composing a detailed defense of Tycho and the true definition of astronomical hypotheses.[5] Kept from fully committing his work to the cosmographical agenda envisioned in the *Mystery*, Kepler continued to probe questions such as the proportion between the speeds of the planets and their distances from the sun. Although Tycho remained reserved about revealing his observations, Kepler harnessed the few details Tycho gave him to explore new ideas, moving the center of planetary motion from the mean sun, a mathematical point, to the real sun.[6] Kepler's tenure under Tycho ended abruptly when the latter died in October 1601.

A week after the death of Tycho, Kepler was summoned as his successor. His responsibilities as Imperial Mathematician consisted mainly in the completion of the *Rudolphine Tables*, together with a number of astrological tasks. Kepler often experienced delays in his salary, forcing him to plead financial hardship in person at the royal palace. As he later reported, the obligation to solicit his salary was something he experienced from the very start of his career. Kepler would spend "two entire months at the royal palace" simply confirming his appointment before waiting another two months to receive his first payment:

> Believe me when I say that I wasted two entire months in attendance at the royal palace. For after the death of Tycho on 24 October, [Johannes] Barvitius declared an imperial salary for me on 26 October and entreaty was necessary to confirm it. I finally received my first payment on 9 March.[7]

Despite salary delays and other difficulties, Kepler's time in Prague is generally considered the most prolific of his career. His work at the court of Rudolph II has been described as his "most serene and most intense twelve years" of activity, favored by a "level of liberty" that rivaled Galileo's

---

[4] For a full-scale study of this dispute, see Jardine and Segonds, 2008.

[5] On Kepler's fulfillment of these two assignments "for his own ends," see Voelkel, 2001, pp. 99–129.

[6] On Kepler's move from a heliostatic system to a heliocentric one, see Gingerich, 1993, pp. 333–339.

[7] *JKGW*, 15, no. 323, 215–220: "Crede mihi quòd duos integros menses stando consumpserim in equestri palatio. Nam mortuo Tychone B. M. 24. Oct. Barwitius 26. Oct. mihi ultrò salarium Caesarium annunciavit; id ut confirmaretur, petendum erat. Donec tandem 9. Marti primam accepi pecuniam." Cf. Voelkel, 2001, p. 142; Caspar, 1993, pp. 122–123.

"happiest period of life in the Republic of Venice."[8] As Imperial Mathematician, Kepler began a new phase in his career fostered by a court culture of novelty. Unlike earlier astronomers who had approached their practice mathematically, Kepler conceived his role as a cosmological craftsman, an "astronomer-philosopher" whose chief purpose was to establish "a true and complete world system."[9] Kepler gave expression to this view in the revolutionary line of research he developed soon after the death of Tycho. In November 1601, he began applying physical reasoning to derive rather than merely describe the eccentricity of Mars.[10] Geometrical devices such as the epicycle were now used to describe the path of planetary motion while physical suppositions such as the motive force of the sun became the distinguishing features of Kepler's causal astronomy.

Before completing his study of Mars in 1605, Kepler would direct his attention to a spectacular celestial phenomenon. In September 1604, a brilliant new luminary appeared in the sky. As motionless as a star, it emerged in close proximity to the conjunction of Mars, Jupiter, and Saturn less than a year after the onset of a major astrological event. This was "the first year of the astrological period of 800 years," Kepler wrote, the eighth such period "since the creation of the world."[11] Made up of four parts, this period began with the 200-year interval known as the Fiery Trigon, which saw Jupiter conjoin with Saturn solely in the fiery signs of Aries, Leo, and Sagittarius. The precise start of the Fiery Trigon had occurred on 7/17 December 1603, Kepler wrote, with "the conjunction of Jupiter and Saturn in 8° Sagittarius."[12] In September 1604, Mars had "moved past Saturn in 10° Sagittarius and in the next few days hurried towards Jupiter, which had already outrun Saturn by a few degrees."[13] The appearance of the new luminary near the three superior planets gave Kepler a chance to make sense of celestial novelty according to his new and causal astronomy.[14] Kepler also took the opportunity to lash out against those astrologers who

---

[8] Bucciantini, 2003, p. 155: "But the freedom in Prague did not last for long. The grave crisis brought on by internal conflict in the House of Hapsburg created a climate of growing uncertainty and concern in the Bohemian capital. Thus, in the autumn of 1610 Kepler expressed his wish to Galileo to move to Padua."

[9] Jardine, 1998, pp. 53–55, 58–59; cf. Westman, 1980, pp. 126, 133–134.

[10] Voelkel, 2001, p. 130.

[11] *JKGW*, 1, 157.8–9.

[12] Ibid., 157.9–11.

[13] Ibid., 157.16–18.

[14] On Kepler's deployment of the new luminary in defense of "a new understanding of cosmic proportion," see Boner, 2011a.

had raced to print fly-by-night predictions as soon after the appearance of the luminary as possible.

For many years, astrologers had eagerly anticipated the Fiery Trigon, predicting "extraordinary prodigies and the commotion of countries ensuing in 1604."[15] When a new luminary suddenly appeared less than a year after the onset of the Fiery Trigon in close proximity to the conjunction of the three superior planets, a virtual flood of predictions poured from the presses.[16] Critical of these predictions along with the general practice of astrology, Kepler described the art as "disgraceful in large part" and weighed down by "great foolishness."[17] His opposition to other astrologers did not stop him from seeing something special in the new luminary, however. To begin with, the time and place of appearance had promised a large and learned audience of observers. The start of the Fiery Trigon had suggested several precious opportunities for the science of the stars, drawing "the eyes of every astronomer" to the very area of the heavens where the luminary appeared:

> [It appeared] when the eyes of every astronomer who diligently attends to his occupation would be directed towards that spectacle of nature during the evenings of those days, some to correct the motions [of the stars], some for pleasure, some to celebrate the return of so many centuries to their starting point, and some for the study of predictions ...[18]

In his interpretation of the new luminary as a star, Kepler stressed that the earth was uniquely in a position to appreciate the proximity of the three

---

15 *JKGW*, 1, 157.12–15.

16 On the "great excitement" aroused by the new luminary in central Europe, see Hotson, 2000, pp. 187–190.

17 Ibid., 165.19–20.

18 Ibid., 157.21–25: "Cùm omnium Astronomorum, qui gnaviter suam professionem obeunt, oculi ad hoc naturae spectaculum, continentium dierum vesperis diligenter essent intenti; alii corrigendorum motuum; alii delectationis; alii solennitatis, tot saeculorum interiectu redeuntis; alii praedictionum studio ..." Kepler also considered the new luminary a call to repent in our search for salvation. There was little coincidence, he claimed, that the proximity of the three superior planets could only be observed from the Earth, where God's message of humility was exclusively intended. See ibid., 291.16–27: "This new celestial prodigy was associated by God Almighty Himself with the three superior planets Saturn, Jupiter and Mars, then conjoined, by a certain deliberation aimed at the salvation of men. For there is nothing in the world, neither big nor small, whose author is not God. And He prefers the human race, living in the limits of this humble lump of earth yet nevertheless His own image, more than any star, even if it exceeded the magnitude of the entire orb [of the Earth] 100,000 times. To mark the place and time of the great conjunction of the three superior planets, as if a monument to remember the event forever and recall the human race to the most remarkable things, He spared no anxiety, no exertion, no exhaustion. In this way, He created something that could be discerned by earth dwellers in the form of such a star."

superior planets. By no accident, he argued, the human race had been put on this "eminent place of habitation," where gifted practitioners could glorify their creator through the institution of astronomical knowledge.[19] The coincidence involved in the appearance of this new star suggested to Kepler a clear sign of divine intervention. Appearing around the time and place of the three superior planets, it provided a salient example of why the earth, whose movement allowed for the measurement of "otherwise inaccessible things," had been set in motion around the sun. Kepler later expressed this view in his *Conversation with the Sidereal Messenger* (1610). There, he suggested that the smallness of the earth belied the divine purpose of our planet, whose annual motion around the sun allowed us to survey the heavens. Our very "cause of knowledge" had called for a moving base of observation, Kepler explained, a purpose that far outshone the meagerness of the earth in comparison with the sun:

> For the sun is in the center of the world, it is the heart of the world, the source of light and heat, and the origin of life and motion in the world ... For although God neither possesses a body nor requires a dwelling place, He expresses in the sun greater virtue governing the world than in any other globe. And so let man acknowledge by the distinction of his dwelling place his own destitution and God's abundance ... Yet the cause of contemplation for which man was made, endowed and equipped with eyes, could not leave him motionless in the center. Rather, the cause of knowledge requires that man be moved round on this earthly vessel in an annual motion. In this way, the measurers of otherwise inaccessible things move from station to station, arriving at a proper foundation by triangulation from the separations of these stations.[20]

Here, Kepler also implied that the heliocentric perspective promised advantages unforeseen in the geocentric view. In this way, heliocentrism confirmed Kepler's anthropocentric outlook.[21]

---

[19] *JKGW*, 4, 309.1–10; cf. Hallyn, 1990, p. 244.

[20] *JKGW*, 4, 308.36–309.10: "Sol quidem in centro mundi est, cor mundi est, fons lucis est, fons caloris, origo vitae motusque mundani est ... Nam etsi Deus corpus non habet nec habitaculo indiget; in Sole tamen ... plus exerit virtutis, quâ mundus gubernatur, quàm in globis caeteris. Agnoscat igitur Homo ipsius etiam habitaculi sui distinctione suam indigentiam, Dei abundantiam ... Adde ... contemplationis causa, ad quam homo factus, oculisque ornatus et instructus est, non potuisse hominem in centro quiescere; sed oportere, ut navigio hoc Telluris, annuo motu circumspacietur, lustrandi causa: non secus atque mensores rerum inaccessarum, stationem statione permutant, ut triangulo mensorio iustam basin ex stationum intervallis concilient."

[21] On the various ways in which Kepler continued to assign "a certain special importance" to the Earth, see Kozhamthadam, 1994, p. 175.

Kepler considered the observational accounts of other astronomers in his study of the new star. Despite their disagreement over the nature of the new luminary, they generally agreed on its physical appearance. As Kepler reported, it was widely regarded as round in shape and emitting a rapid scintillation. These features were interpreted by Kepler—and apparently "everyone who witnessed the first appearance" of the new luminary—as the characteristics of a star situated in the outer sphere of the cosmos:

> Everyone who witnessed the first appearance of the star agrees that it was perfectly round, with neither hair nor a beard nor a trailing robe extending in any direction, for there was no appearance of locks of hair nor a bearded or tailed comet. Rather, it was similar to the fixed stars, with rays of light leaping in every direction like [those] of the fixed stars. It was of an extremely clear scintillation, with a gleaming or glittering so swift that some would deny to themselves as long as they lived that anything ever seen in the heavens was of an equally brisk motion.[22]

Kepler also reported that the new star became redder as it approached the western horizon. This change in color had led some astronomers to liken it to a torch, he wrote, displaying a "continuation of flames interrupted by the impulse of winds."[23] Given the remarkable range in radiance of the new luminary, however, Kepler preferred to compare it to a diamond. The initial luminosity was so intense, in fact, that it outshone even fixed stars of the first magnitude as well as Saturn and Mars. The calm and coaxing light of Jupiter, also outshone by the nearby luminary, could be "easily distinguished from the turbulent glimmering of the star for all of October."[24]

Contrary to the majority of authors who rushed to publish as soon as possible after the appearance of the new star, Kepler completed his own study, *On the New Star*, in September 1606. In his dedicatory preface to Rudolph II, Kepler explained that he had waited to give a full account of the new luminary until after it had disappeared in February 1606. Unlike those who had written more hastily, Kepler claimed his delay had allowed him to strengthen his account through "astronomical, physical,

---

[22] *JKGW*, 1, 160.18–23: "... convenit omnibus, qui primo eius exortui advigilarunt; fuisse exactè rotundam, nullo crine, nulla barba, vel syrmate in ullam partem projecto, quare nulli Crinitarum speciei, neque Pogoniae neque Cometis accensendam; sed stellis fixis similimam, radiis undiquaque ut fixarum, emicantibus; scintillatione clarissima, coruscatione seu vibratione tam rapida, ut negaverint quidam, sibi dum viverent, unquam quicquam in coelo visum esse aequè pernici motu ..."

[23] Ibid., 160.28–29.

[24] Ibid., 161.4–7.

and metaphysical disputation."[25] Armed with this advantage, Kepler had finished the first part of his book just in time for the Frankfurt Book Fair. He compared his "highly intricate and variously hindered occupation" in astronomy with Rudolph's "vast and ongoing war with the Ottoman Empire."[26]

The Multiple Purposes of On the New Star

Although it has been described as one of Kepler's "more quirky and much less 'modern'" works,[27] *On the New Star* represents a central part of Kepler's larger cosmological program. Among his main objectives, Kepler set out to show how the location of the new luminary in the sphere of the fixed stars offered further support for the heliocentric system. In opposition to Tycho, who disapproved of the enormous distance Copernicus had suggested between Saturn and the sphere of the fixed stars, Kepler argued that the remarkable remoteness of the new luminary was proven by an absence of parallax.[28] For his own part, Tycho had placed the sphere of the fixed stars at twice the distance of Saturn, "a form of the world" that Kepler called into question.[29] It made little sense, Kepler wrote, to measure "the immense multitude of the fixed stars" according to a proportion "with the star of Saturn alone."[30] Kepler also challenged the widely held view that the sphere of the fixed stars revolved at such a swift rate. On the contrary, he claimed that the new star suggested their immobility. This proved to be particularly lethal to those who located the luminary below Saturn in one of the solid celestial spheres moving the planets. If the star was below Saturn, Kepler argued, an absence of parallax could only be produced by the continual penetration of the orbs. According to the prevailing view of Ptolemaic astronomy, however, the penetration of the orbs was entirely impossible:

---

[25] Ibid., 151.26–30.

[26] Ibid., 152.6–10.

[27] Gingerich, 2002, p. 237. On the other hand, Gingerich has also claimed that "any complete assessment of Kepler's place in astronomy would necessarily notice his *De stella nova* (1606)." See Gingerich, 1993, p. 331.

[28] *JKGW*, 1, 235.28–35. On Kepler's reading of this remoteness as a sublime proportion, see Boner, 2011a.

[29] *JKGW*, 1, 235.37–39.

[30] Ibid., 236.1.

> [Ptolemy] considers all of the orbs solid, from the innermost moon to Saturn farthest away. Thus, if the star were within that region of the heavens of Saturn (since it is all but certain according to all three forms of hypotheses that it is located above the moon), it would therefore be in one of the solid orbs. And since the penetration of solids is not permitted, it would be carried along with the orbs, since none of them remains inert and immobile.[31]

Along with taking to task the two rival hypotheses of Ptolemy and Tycho,[32] Kepler turned to the reform of astrology in *On the New Star*. Echoing his earlier letters to Herwart, Kepler declared the aspects "practically the only thing in astrology worth preserving."[33] Kepler devoted eight straight chapters to separating this kernel of astrology from the surrounding husk before ever attempting to explain the significance of the new star. The illustration he placed on the frontispiece of his book vividly captured this approach to astrology. Surrounded by small chicks, an eager hen "searches through the muck and yields grains." It is tempting to interpret the image of the hen as Kepler, searching for seeds of truth among his less scrupulous contemporaries.

Kepler envisioned his dual role in *On the New Star* as both a defender and reformer of astrology. He measured these roles according to his master and predecessor, Giovanni Pico della Mirandola (1463–1494), author of the *Disputations against Divinatory Astrology* (1496), "the most extensive and incisive attack on astrology the world had yet seen."[34] Kepler devoted Chapters 7, 8, 9, and 10 of *On the New Star* to defending astrology in the light of Pico's daunting critique. Those parts of astrology that Kepler chose

---

[31] Ibid., 238.33–239.5: ". . . habet ille omnes orbes solidos, inde à Luna infima, usque ad Saturnum extimum. Igitur, si stella esset intra complexum coeli Saturnii (quia per omnes tres hypothesium formas certissimum est, supra Lunam collocari), esset igitur in aliquo orbium solidorum. Ac cùm non detur penetratio corporum, planè raperetur cum orbibus; quia eorum nullus est, qui stet iners et immobilis." On the solidity of the celestial orbs in Aristotle, Ptolemy, and the medieval scholastic tradition, see Barker, 2011; Granada, 2006; Lerner, 1989; Rosen, 1987.

[32] Bucciantini, 2003, p. 139: "Like all of his other works, *De stella nova* was a Copernican text. It was not, and could not have been, a Copernican work in the strict sense, containing some explicit attempt to demonstrate from the new celestial phenomenon the 'truth' of the heliocentric system. Nevertheless, Copernican ideas are recalled and repeatedly discussed, as in Chapter 15, where the immobility and immense distance of the new star from the center of the Earth are demonstrated "in the Copernican hypothesis" according to the calculation of annual parallax, or elsewhere in Chapter 21, where Kepler appeals to Copernicus and goes on guard against anyone who (referring especially to the "miserable" Bruno), abusing the authority of the great Polish astronomer, becomes the supporter of the insane conception of the infinitude of the world."

[33] *JKGW*, 1, 166.40–167.1.

[34] Vanden Broecke, 2003, p. 55.

Figure 2. Frontispiece of *On the New Star* (1606). Johannes Kepler, *De stella nova in pede Serpentarii* (Prague, 1606). Courtesy of the Dibner Library of the History of Science and Technology, Smithsonian Library Institutions.

to preserve he portrayed as "still not snatched away" by Pico,[35] while those parts he rejected were backed by the same arguments Pico had deployed in his ruthless campaign. On either front, Kepler's "running debate" with Pico allowed him to reform astrology according to an established tradition.[36] Against this backdrop, Kepler refined his arguments in favor of the few elements he accepted. In this way, he spared himself the task of repeating the work of his learned predecessor, to whom Kepler conceded "everything concerning the foolishness of astrology."[37]

In his dialogue with Pico, Kepler distinguished between the natural and what he considered the purely cultural.[38] Human artifacts such as the twelve-part division of the zodiac served practical purposes that Kepler claimed had nothing to do with the natural world. Even the four elements were only indirectly related to the heavens. "The stars do not dry or moisten inferior things of their own accord," Kepler wrote, "but accidentally."[39] Kepler further argued that the four elements had even less to do with the zodiac, having been adopted without any regard for "the nature of things" in the first place.[40] "Why are the three signs of Gemini, Cancer, and Leo not fiery," Kepler asked, "since they signal the summer? And why are the vernal signs not aerial?"[41] Kepler applied this same line of critical reasoning to the Fiery Trigon. Despite its rarity and utility as a chronological reference, it was classified by Kepler as just another object of convenience.

The start of the Fiery Trigon in 1603 signaled a return to what Kepler thought might have been the original configuration of the heavens, a

---

[35] *JKGW*, 1, 181.12–13. For a comparison of the two views of Pico and Kepler on astrology, see Rabin, 1987.

[36] On Kepler's "running debate" with Pico, see Westman, 2001, p. 230.

[37] *JKGW*, 1, 184.13–16. One element of astrology that Kepler opposed along with Pico was the assignment of evil properties to the heavens. In particular, Kepler rejected the role of the heavens in events such as "the terrible changes experienced in the Christian religion." See Vernet, 1972, p. 459. For a full translation of Chapters 7, 8, and 9 of *On the New Star*, see Boner, 2010.

[38] On this distinction in Kepler's astrology, see Simon, 1975, pp. 444–445.

[39] *JKGW*, 1, 178.3–5.

[40] Ibid., 165.33. In the *Disputations*, Pico condemned the causes given by the astrologers "for the trigons as well as the gender of the astrological signs." Denominations according to the elements, such as the 'fiery' signs of Aries, Leo, and Sagittarius, were found "to lack reason," while Pico suggested that associations of gender with the signs of the zodiac were better suited for "ridicule rather than serious consideration." Pico criticized the signs themselves as "contrivances of the mathematical astronomers for the facilitation of calculation." See Garin, 1952, pp. 36, 110, 114.

[41] *JKGW*, 1, 178.8–9.

regression, so to speak, of 5,600 years. The recurrence of the Fiery Trigon every 800 years divided the history of the world into seven broad periods. Kepler represented each of these periods by a person of renown and a select few "coinciding things," cautioning the reader that they were not "the effects of the Trigons" but purely coincidental.[42] The first Fiery Trigon marked the age of Adam and "the creation of the world," the sixth the birth of Christ and "the reformation of the world," the seventh the reign of Charlemagne and "the Christian and Islamic empires," and the eighth and most recent the rule of Rudolph II.[43] The ninth Fiery Trigon, anticipated fully 6,400 years after the first one, signaled an age of uncertainty for Kepler. "What will become of our prosperous Germany," he asked, "and who will be our successors?"[44] While Kepler remained willing to accept the Fiery Trigon as the basis of such a chronological compendium, he saw it as little more than a useful device "for aiding the memory."[45] He explicitly rejected the determination of empires and religions according to the conjunctions of Jupiter and Saturn as "foolish and superstitious."[46]

Yet Kepler went further in deposing the doctrine of the Fiery Trigon. He first showed that it did not recur at precisely the same interval of time and could slip into another sign. Cardano had calculated that the Fiery Trigon would actually begin in 1583, Kepler reported, when the average conjunction of Jupiter and Saturn would first enter the fiery sign of Aries.[47] Calculations made by Kepler, on the other hand, suggested that while the two superior planets had conjoined in 8° Sagittarius in 1603, they would conjoin again in the aqueous sign of Pisces in 1643. This would break the cycle and reveal an irregularity that Kepler saw as a telltale sign of an artificial thing:

> For I have said that the Fiery Trigon marks the time in which no conjunction of the superior planets occurs except in the fiery [signs]. And although this is the case in the years 1603 and 1623, it varies in the year 1643. At that time, while the average conjunction [is] in Aries, the actual one occurs in 26° Pisces, with retrogradation carrying the planets backwards.[48]

---

42 Ibid., 183.1–3.
43 Ibid., 183.4–16.
44 Ibid., 183.17–21.
45 Ibid., 182.23–25.
46 Ibid., 188.41–189.3.
47 Ibid., 182.33–34.
48 Ibid., 183.28–32: "Dictum enim, tempus id vindicari igneo Trigono, intra quod congressus superiorum nulli fiant, nisi in igneis. Atqui hoc, etsi anno 1603. 1623. sic habet, anno

How could the conjunction of Jupiter and Saturn suffer such a relapse? Although the angular separation between two consecutive conjunctions was ordinarily 117°, it could end up being as little as 102°. This had occurred between 1583 and 1603, for example, when the conjunction had only moved from 21° Pisces to 8° Sagittarius. Such a separation was "closer to a square," Kepler wrote, "than to a triangle."[49]

Kepler's critical study of the Fiery Trigon served to deepen his distinction between the artificial and natural worlds. His development of this division has been attributed in part to Pico, whose emphasis on the physical nature of the heavens encouraged Kepler to identify how our knowledge of the stars was obscured by our own mythology.[50] In his great assault on astrology, Pico had limited the influence of the heavens to the heat, light, and motion arriving at the sublunar level. Pico had outright rejected the idea that Jupiter and Saturn possessed "more power together in the great conjunctions than apart," since the intensity of their light would never surpass what they reflected separately.[51] On the contrary, Pico claimed that the obscuration of one of the planets would cause something lesser, if anything, than if they operated individually. In the end, Pico dismissed the doctrine of the great conjunctions as "a recent invention, born from a misunderstanding of Ptolemy," whom Pico accused of rashly interpreting the relevant sources.[52] Although every major event appeared to follow a great conjunction, there was no real source of influence, Pico argued, only a point of reference that conveniently occurred every twenty years. "There has never occurred any great legal or political change that some conjunction of the superior planets did not precede," Pico wrote, a simple fact that also applied to "changes in lesser things."[53] Pico saw this as fodder for fallacy, as astrologers continued to draw connections with the heavens that did not actually exist.

In his reply to Pico, Kepler suggested a more sublime source of influence than heat, light, and motion. The light of the sun, arriving directly and indirectly from the surfaces of the other celestial bodies, was clearly a vehicle of influence for Kepler, who suggested that it arrived in various

---

tamen 1643. variat. Tunc enim media conjunctio in Ariete, vera autem in 26. Piscium est, retrogradatione Planetas in anteriora trahente."

49 Ibid., 183.38–184.3.

50 See Rabin, 1987, p. 58; Westman, 2011, pp. 320–321.

51 Garin, 1946, 544.

52 Ibid., p. 550.

53 Ibid., p. 558.

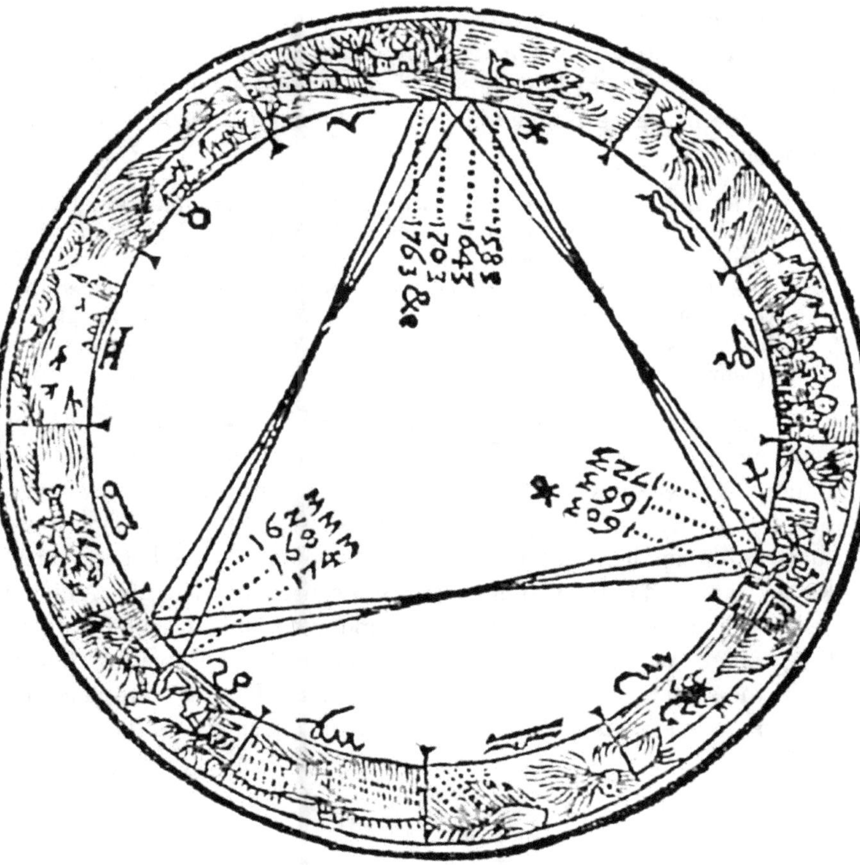

Figure 3. Great conjunctions, 1583–1763. Johannes Kepler, *De stella nova in pede Serpentarii* (Prague, 1606), p. 25. Courtesy of the Dibner Library of the History of Science and Technology, Smithsonian Library Institutions.

mixtures.[54] The archetypal cause of the aspects, however, took pride of place in Kepler's astrology.[55] These configurations produced by the positions of the planets stimulated responses from the biosphere of living beings inhabiting the sky, surface, and interior of the earth. In particular, Kepler underscored the difference between the passive reception of the physical properties of heat, light, and motion and the active ability of a sublunar soul to identify the archetypes in the appearance of the heavens. Kepler claimed that this explained why the great conjunctions of Jupiter and Saturn exerted any influence at all. As he pointed out in his response to Pico, this influence ultimately lay with "sublunar nature," not the planets themselves:

> ... Pico asks why one should suppose that Saturn and Jupiter bring about greater things when they are together than when they are apart. I will respond according to my own conviction and not that of the astrologers: the work that we assign to the superior planets when they are conjoined that we do not apply when they are apart by no means comes from the planets themselves (save for [their] illumination and calefaction alone), but from sublunar nature [*natura sublunaris*].[56]

The planets did not exert an influence like "natural agents," Kepler explained, moving the earth "through the presence of some sensible thing."[57] Rather, they acted the same way on earth as "light the eyes, sound the sense of hearing, and heat the sense of touch," though more subtly. Although light rays from every celestial body arrived at the earth, Kepler claimed that their configuration was sensed instinctually. "A divine instinct" allowed the soul of the earth and every human being to identify geometrical principles in the appearance of the heavens.[58] Essentially, the same principles involved in our creation were exemplified by certain arrangements of the heavens, spurring a response to our celestial

[54] *JKGW*, 1, 187.25–26.

[55] On the complementarity of light and archetypal causes in Kepler's early astrology, see Westman, 2001, p. 232.

[56] *JKGW*, 1, 184.23–28: "Quaerit Picus ... quare credit, maiora facere Saturnum et Jovem, cùm sunt simul, quàm cùm sunt separati? Respondebo ex mea, non ex Astrologorum sententia: Opus quod superioribus adscribimus junctis, quod non competit separatis; id nequaquam Planetarum ipsorum est (praeter nudam illuminationem et calefactionem), sed ipsius naturae sublunaris."

[57] Ibid., 184.29–31.

[58] Ibid., 269.6.

surroundings. In this way, the soul was born with the ability to relate the positions of the sun, moon, and planets geometrically.[59]

In opposition to those whom Kepler criticized for measuring sublunar nature "with a short stick," he pictured the earth as the dwelling place of a soul whose abilities went beyond those of human beings.[60] This soul became aware of configurations in the heavens that every so often assumed the form of a figure that embodied the same principles originally imprinted on the soul. Kepler considered this sudden correspondence of external appearances with the internal archetypal principles of the soul a reawakening. He would later describe the essence of harmony in the *Harmony of the World* similarly, as the outward expression of an archetype of harmony in the human soul. There, Kepler would suggest that our souls identified "order and proportion" in sensible harmonies by reference to their own archetypal principles.[61] "To find a suitable proportion in sensible things," Kepler wrote, "is to uncover and recognize and bring to light a similarity of that proportion in sensible things with some particular archetype that lies within the soul."[62] Pico had denied this sort of ability to the earth along with several others, assuming that they surpassed even "those faculties that man possesses."[63] In his response, Kepler gave an everyday example that showed how some human faculties were excelled by those of other, allegedly lesser living beings:

> Tell me, Pico, with what sense does a dog perceive the traces of his master—by smell? And yet man's sense of smell is not so sharp! Thus, you discover that there is something in the dog that you had not discovered in man. Nor would you have believed those things about a dog unless you had more often approached the matter up-close yourself.[64]

Kepler made his point clear. If the faculties of man could be outdone by a dog, was there any reason why the soul of the earth should not also

---

[59] On the "geometrical instinct" that led the sublunar soul to sense "the conjunctions of the stars and the harmonies of the world," see Mahnke, 1937, pp. 138–139.

[60] *JKGW*, 1, 185.17–20.

[61] *JKGW*, 6, 215.36–39.

[62] Ibid., 215.30–33. On Kepler's later theory of harmony "as a product of the soul," see Escobar, 2008, p. 31.

[63] *JKGW*, 1, 185.17–20.

[64] Ibid., 185.22–25: "Dic mihi Pice, quo sensu canis vestigia domini percipit: num odoratu? At hercle hominis tam exquisitus odoratus non est. Aliquid igitur discis in cane, quod non didiceras in homine; nec credidisses de cane referentibus, nisi in rem praesentem saepius ipse venisses."

surpass human beings in some ways? As his parting shot, Kepler claimed that Pico would have known this had he ever taken the time to observe his immediate surroundings.

Kepler attributed to the soul of the earth an original impression of archetypal principles that allowed it to identify their external appearance in the heavens. Kepler argued that these archetypes were part of a "spiritual formative faculty [*spiritalis facultas formatrix*], which others refer to as seminal reason [*ratio seminarium*]."[65] Every living being preserved an impression of the archetypes, Kepler said, through this formative faculty. To explain this more clearly, Kepler turned to the analogy of certain plants that continue to display colors after originally absorbing them. Why, he asked, did the flowers of these plants adopt the colors conveyed by their roots long after these colors were no longer in supply? How did drops of color appear on one flower, while on another the color was completely absorbed? And why was the color not digested in such a way that it simply accrued to the body of the plant like every other property of the fluid? In the face of these mysteries, Kepler favored the formative faculty as a way of making sense of them.

Rather than "crawl humbly through plants,"[66] however, Kepler found a second example of the formative faculty in the Bible. There, he cited the story of Jacob and his pledge to claim possession of only those sheep that were not white. To ensure that the population of "speckled and spotted" sheep grew greatly, Jacob placed rods of poplar in the watering troughs where the flocks came to drink. These rods had been peeled to give the appearance of stripes, which would produce multi-colored offspring in the sheep that saw them while copulating near the troughs. Kepler read this passage as proof of the impression of external appearances on the formative faculty and the effect this could have on the conception of the offspring:

> Observe the pregnant sheep of Jacob the Patriarch. With the lamb having witnessed several [striped] rods beneath the clear waters, it manifested the appearance in the offspring. How this appearance would come from the

---

[65] Ibid., 185.35–37. Kepler's account of seminal reason recalls that of Marsilio Ficino (1433–1499), who identified it as the efficient principle of form for every living species. For more on these "reason-principles" in Ficino's view of the vegetative power of the World Soul, see Hirai, 2002.

[66] *JKGW*, 1, 185.38. With his reference to crawling "humbly through plants," Kepler may have had Pliny in mind. On Pliny's self-portrayal as "a man of the earth" and his emphasis on the close connection between agriculture and astronomy, see Taub, 2003, p. 173.

> plant rod to the eye, I do not inquire, as it is common knowledge. But how [it would come] from the eye into the womb, into the formative faculty, and on into the fetus, how, I say, from this plant rod by that angle—that requires explication.[67]

As the experience of Jacob affirmed, the formative faculty altered the appearance of the offspring in response to external events. Kepler suggested that the conception of certain weather conditions was moved similarly by the influence of the heavens on the soul of the earth.

The Soul of the Earth: Instinctual Responses to the Astrological Aspects

In his study of the effects of the new star on the weather, Kepler pictured the earth as an anatomical system governed by a soul. This account often employed the ideas of ancient authorities such as Pliny, whom Kepler praised for his aversion to "astrological superstition,"[68] and Seneca, whose *Natural Questions* Kepler had not read before writing the *Cosmographical Mystery*.[69] Kepler later explained in the *Harmony of the World* that he simply observed that "all of the many things coming from the body of a living being that bear witness that there is a soul in it also come from the body of the earth."[70] It is clear, however, that Kepler's conception of the earth was also informed by the accounts of these ancient authorities. The line between anatomical analogy and reality is difficult to discern in their works,[71] and the same can be said for Kepler. Water was portrayed as a form of provision and perspiration for the body of the earth. Springs and rivers were the outpourings of internal networks of "aqueous veins," while geysers were the discharges of "digested vapors" that emerged in

[67] *JKGW*, 1, 185.38–186.2: "Sed quid ego humilis inter plantas repo? Praegnantem aspice ovem Jacobi Patriarchae. Variegatos illa baculos sub aquis limpidis contuita, speciem in foetum derivavit. Quomodo species haec à baculo in oculum venerit, non quaero; vulgare est. At quomodo ab oculo in uterum in formatricem facultatem, in foetum, quomodo, inquam, ab hoc baculo ad illum angulum: hoc eget explicatione."

[68] Ibid., 165.26–28. Kepler cited Pliny as a source of support when locating the new luminary in the sphere of the fixed stars.

[69] *JKGW*, 8, 22.11–13.

[70] *JKGW*, 6, 268.36–37.

[71] As Taub explains in her survey of these analogies in ancient texts, "[Seneca's] drawing out of the analogy between the earth and the human body is elaborate and vivid, and more detailed than that found in Aristotle's *Meteorology* or Manilius's *Astronomica*." See Taub, 2003, p. 144.

the same way as sweat."[72] The watery conduits coursing through the terrestrial interior and covering the surface were controlled by the soul of the earth, Kepler claimed, swallowing sea water "into hidden sites of digestion" and expelling vaporous fumes in the form of clouds, fog, mist, rain, and other aqueous substances.[73] While these everyday activities followed a general ebb-and-flow, the appearance of an aspect in the heavens could supply a source of great acceleration.[74] Never one to exclude geographical differences from the determination of these changes, Kepler conditioned their consequences on regional characteristics. At the onset of an aspect, coastal areas might suffer flooding, northern parts might experience sleet or snow, and dry regions might suddenly receive rain.[75]

How did these various outpourings suddenly intensify with the inception of an influential configuration? Kepler relied on another anatomical analogy to resolve this question. He supposed a faculty in the earth that expelled vapors in the same way that animals produced "a seminal excretion."[76] The soul of the earth appeared to react to the aspects by spurring the normal activity of this faculty and fuelling the ejaculation of fluid. Kepler suggested that our planet might even derive some pleasure from this process. "It is not so absurd, really, that we should attach some pleasure to this excretion," he wrote, "since the earth shares so many other things with animals."[77] With the inner vessels of the earth swelling at the appearance of an aspect, the emission of fluid accompanied this poignant impression with a certain sense of relief:

> In fact, physicians tell us that if the blood vessels swell with seminal fluid when a pleasant image is presented during sleep, an excretion occurs at once without any contact taking place. And is there anything more similar to such a thing than what is known to happen with that faculty in the earth, the perceiver of celestial aspects that exudes rainy vapors when stimulated by one of these aspects? On the one hand, the immaterial image, the object

[72] *JKGW*, 1, 317.13–25. Cf. Pliny, *Natural History*, 2.166–170. In the *Natural Questions*, Seneca viewed subterranean humors as the earthly equivalents of "brain in the head, marrow in the bones, mucus, saliva, and tears." And when comparing the earth's underground channels to veins and arteries, Seneca described rivers and streams as incisions on the surface of the Earth flowing until the blood was "exhausted" or the cut was "closed up." See Seneca, *Natural Questions*, 3.15–16.

[73] *JKGW*, 1, 317.13–15.

[74] Ibid., 317.20–21.

[75] On Kepler's attention to geographical differences in his astrometeorology, see Rabin, 1997, pp. 750–751.

[76] *JKGW*, 1, 317.15–16.

[77] Ibid., 317.21–23.

> of the imagination, rouses the reproductive substance, so what, on the other hand, prevents the existence of pleasure in the expulsion of that substance from the perception of this image?[78]

Kepler compared this analogy to Virgil's image of spring in the *Georgics*, where the "fertile showers" of the "sovereign father of heaven descend to the womb of his joyful spouse" in the terrestrial interior.[79] In contrast with Virgil, Kepler claimed that these showers were the creation of a soul who actively responded to the heavens rather than simply following their cue. "I attribute nothing to the heavens," he wrote, "except for that venereal image, the aspect, provoking the outflow of this seminal fluid from the earth."[80]

As with the aspects, the appearance of the new star was not a straightforward source of influence for Kepler. Remarkably brilliant, it was originally as bright as two or three other stars and soon displayed an appearance "similar to the Dog Star or the Shoulder of Orion."[81] Apart from a certain level of heat and light, however, the influence of the new luminary involved the active participation of a sublunar soul. The heavens held no direct form of influence "except for illumination and warmth," Kepler wrote, and he focused rather on their reception by "sublunar nature."[82] Kepler claimed that every form of life on earth had sensed the sudden appearance of the star. Nothing happened in the heavens that was not felt "by some hidden cause" on earth, he said.[83] Kepler also suggested that the star had stirred these souls in the same way as any other celestial novelty. "Whenever something new and extraordinary appears in the heavens," he wrote, "sublunar nature trembles in some way."[84] Given the remarkable circumstances surrounding the star, however, Kepler was inclined to believe that it bore a special significance.

---

78 Ibid., 317.23–30: "Etenim docent Medici, si quando humore genitali tument venae, facile vel per somnum, imagine dulci objecta, nullo etiam contactu accedente, fieri excretionem. Quid vero huius rei similius, quam quod constat inesse in Terra facultatem, aspectuum coelestium perceptricem, quae stimulate aliquo aspectu, exsudet vapores pluvios? Utrinque species immateriata, objecta phantasiae, ciet materiam genitabilem: quid impediet igitur utrinque ex perceptione speciei, et expulsione materiae existere voluptatem?" For a survey of the many ways in which scholars have rendered Kepler's use of the term 'species,' see Rabin, 2005, pp. 49–50.

79 *JKGW*, 1, 317.32–35.

80 Ibid., 317.37–318.2.

81 Ibid., 314.31–34.

82 Ibid., 314.29–35.

83 Ibid., 315.19–20.

84 Ibid., 315.21–24.

Beginning with "the effects of the star on the weather,"[85] Kepler first observed a sudden increase in precipitation, particularly when the new star formed part of an aspect. "The winter of 1604–1605 was very wet," Kepler wrote, "especially on those days when the planets were conjoined or configured with the new star."[86] The years following the great conjunction had been fruitful, and the cheap price of wine had returned to roughly the same rate it had been after the previous conjunction in Pisces. The earth had witnessed high levels of humidity bursting from an interior that had become "swollen with fluid."[87] Considering the amount of fluid required to produce such an outpouring, Kepler supposed that the great conjunction had served as a stimulus of the earth to absorb it.[88] Moved in this way, the soul of the earth would thus have sensed the new star near the superior planets as an extraordinary impetus. Kepler suggested that the great thirst of the earth would soon have become a growing burden, however, with the new luminary serving as a welcome source of relief. "Whenever as it was configured with the planets," Kepler wrote, the star "provided a stimulus to the expulsive faculty."[89]

The author of cold and watery vapors, the soul of the earth also produced hot and dry exhalations. Baked beneath the surface of the earth and expelled through fissures and other openings, these exhalations could produce earthquakes, lightning, thunder, and strong winds. Kepler described them as "brazen, fiery, sharp, and sulfurous," a source of "drought and squalor."[90] He compared their sterility to "empty and painstaking intercourse,"[91] a fruitless alternative to the fertile vapors that fostered growth. They were especially significant for weather predictions in the years following the Fiery Trigon, Kepler wrote, since astrologers had "struck fear in the melancholic man of the conflagration of the world."[92] Despite signs of dryness and heat during parts of 1604 and 1605, however, Kepler suggested that the summer of 1606, "which exceeded the average in moisture," "completely shattered" this troubling prediction.[93] He recalled similarly contradictory results for astrologers in 1524, when drought and

85 Ibid., 316.21–22.
86 Ibid., 318.31–35.
87 Ibid., 318.35–37.
88 Ibid., 318.38–39.
89 Ibid., 319.13–15.
90 Ibid., 318.11–13.
91 Ibid., 318.13–14.
92 Ibid., 325.18–19.
93 Ibid., 325.20–22.

famine followed the prediction of horrible floods brought on by the Aqueous Trigon. "With no rain that year and the blood of farmers soon in abundant supply," Kepler wrote, "many lost their lives."[94] The lack of any link between the activity of the earth and the elemental denomination of the heavens led Kepler to lament "the infinite foolishness of astrologers." He had made this clear in Chapter 6 of *On the New Star*, where he had revealed this denomination as a clever device that had little to do with nature:

> The expectation of drought and squalor strictly from the sign of the Fiery Trigon is empty. For there shall be no less rain, snow, and flooding during these 200 years than the fire, lightning, and drought that occurred during the Aqueous Trigon [the previous 200 years]. In Chapter 6, I wrote that denominations according to the elements—and so also from fire—did not come to the trigons from nature herself, much less from their effect, but from the sheer judgment of astrologers. Oh the infinite foolishness of astrologers, who never begin to gain knowledge and never cease to found their predictions on those futile foundations of the ridiculous denominations![95]

In his conception of the earth as the creator of wet and dry exhalations, Kepler drew directly from Aristotle and the *Meteorology*. Kepler's view of the generation of metals and minerals also stemmed from the system of exhalations set out by the Stagyrite. According to Aristotle, the dry exhalation produced minerals such as sulfur beneath the surface of the earth. Described as "colored dust or stone consisting of a similar composition,"[96] minerals derived their dryness and inflammability from the solid excrement of dry exhalations enclosed in underground caverns for long periods of time. Metals, on the other hand, were identified as mixtures of the aqueous exhalation condensed beneath the surface with different amounts of the igneous exhalation. Classified according to the amount of igneous exhalation they contained, some metals were lower or less refined than others. While Kepler largely accepted this account of metals and minerals,

94 Ibid., 325.25–26.

95 Ibid., 325.26–34: "Vana est ... squalorum et siccitatum, ex solo nomine ignei Trigoni expectatio. Non minus enim pluviarum, nivium, diluviorum erit per hos ducentos annos, quam exacto aqueo Trigono fuit incendiorum, fulminum, siccitatisque. Dictum enim capite VI. Denominationes ab Elementis et sic etiam ab igne, venisse trigonis non ex Natura sua, multo minus ex effectu; sed ex mero Astrologorum arbitrio. O vanitatem infinitam Astrologorum, qui nunquam sapere incipiunt, nunquam cessant his futilissimis denominationum ludicrarum fundamentis prognostica sua superstruere!"

96 Eichholz, 1949, pp. 144–145. On the authority of Aristotle in early modern meteorology, see Martin, 2011.

their generation in the earth indicated to him the activity of a soul. The same source of rain and rivers, "devouring and digesting sea water" and expelling it "on high," was a soul responsible for producing a whole range of subterranean marvels.[97]

By supposing a soul in the earth, Kepler differed with Aristotle in at least two important ways. First, Kepler granted the earth greater autonomy. While Aristotle had assigned the sun as the cause of the earth's system of exhalations, Kepler suggested a soul whose activity was only supported by the sun. In a later account of the earth's ability to produce life, Kepler criticized those who had granted the sun absolute power on the sublunar sphere "without axes, burins, chisels, or any other material instruments."[98] The task of making use of the heat and light of the sun fell squarely on the earth, Kepler said, and he suggested "a subterranean stronghold" as responsible for redirecting solar resources to the creation and continuation of life.[99] Kepler would later return to Virgil's *Georgics* to compare his account of the earth to the image of "a joyful wife."[100] The soul of the earth collaborated with the sun, Kepler claimed, in the same way that "a wife perceives what is happening to her with pleasure and helps her husband with just the right motion."[101] Employing his own analogy, Kepler wrote that the sun would never reach an agreement with the earth without the cooperation of a soul that would "conspire with the enemy and open the gates to him."[102]

Kepler established further distance from Aristotle by supposing that the soul of the earth possessed the same archetypal principles stamped on the souls of human beings. Rather than focus on the physical interaction of the celestial and sublunar spheres,[103] Kepler gave greater importance to their formal communication expressed in the language of the geometrical archetypes. In a clear break with Aristotelian cosmology, Kepler assumed that these principles were employed in the act of creation and continued to inform the soul of the earth. Consisting of the same constructible polygons underlying the astrological aspects, these archetypes were also accountable for the appearance of "remarkable geometrical figures"

[97] *JKGW*, 1, 268.11–16.
[98] *JKGW*, 6, 266.7–8.
[99] Ibid., 266.14.
[100] Ibid., 266.10–11.
[101] Ibid., 266.11–12.
[102] Ibid., 266.14–16.
[103] On the interaction of these two spheres in Aristotle's cosmos, see Taub, 2003, p. 87.

beneath the surface of the earth.[104] Kepler compared the earth's "hidden impression" of the archetypes to "the imagination of living beings."[105] He added to this impression the earth's ability to realize the archetypes in the form of physical things, echoing in this way the original rudiments of creation. In his preface to *On the New Star*, for example, Kepler had found in gems and minerals the "rudiments of the regular polyhedra," principles that were preserved "among precious things" and aspired to "with the painter's brush."[106] That these things expressed the archetypes so closely gave Kepler cause to commend their beauty and refinement in any form he found them naturally occurring on earth.

Finding Middle Ground: The Soul of the Earth and the Surrounding Cosmos

Kepler identified the regular polyhedra in many subterranean substances. Diamonds consist of combinations of cubes and octahedra, for example, while quartz is composed completely of tetrahedra. Kepler suggested the soul of the earth as the source of such striking configurations. Yet how did he claim it could act as their efficient cause? Let us take a closer look at the nature of the regular polyhedra and their place in Kepler's view of the earth and heavens.

The regular polyhedra remained the scaffolding of Kepler's system of planetary motion, the foundation for his more mature theory in the *Harmony of the World*.[107] As Kepler recalled in his preface to the *Cosmographical Mystery*, he had discovered their role in the relative disposition of the planets while demonstrating the angular distances involved in the succession of great conjunctions. Normally separated by intervals of 117°, consecutive sites of conjunction became the vertices of triangles, or quasi-triangles, which would together form another circle half the diameter of the original one. With one circle circumscribing the triangles and the other circle inscribed inside, the ratio between their diameters recalled the same 1:2 ratio between the dimensions of the paths of Jupiter and

104 *JKGW*, 1, 268.14–16.
105 Ibid.
106 Ibid., 152.27–30.
107 On the enduring importance of the regular polyhedra in Kepler's harmonic astronomy, see Stephenson, 1994, p. 89. Barker describes the polyhedra as the apparatus of "the providential order" Kepler uncovered in the *Cosmographical Mystery* whose "causal structure" he later established in the *New Astronomy*. See Barker, 2004, pp. 157–158, 176.

Saturn. Kepler saw this relation as the first step towards discovering the role of the regular polyhedra. It was an answer to his prayer that "if Copernicus had spoken the truth, things should proceed in this way," and a direct result of divine intervention.[108] Kepler wrote that he had taken this step while teaching his students "the leaps of the great conjunctions" through the zodiac and how they determined the four trigons:

> And so on 9/19 July 1595, when I was about to show my students the leaps of the great conjunctions through eight signs at a time and how they proceed step-by-step from one trigon to the next, I inscribed several triangles, or quasi-triangles, in the same circle, in such a way that the end of one would be the beginning of another. And thus a smaller circle was formed by the points at which the sides of the triangles intersected each other. For the radius of a circle inscribed in a triangle is half the radius of the circumscribed [circle]. The proportion between the two circles appeared to the eye almost the same as what exists between Saturn and Jupiter.[109]

Beginning with the triangle, Kepler applied polygons to the distances between different pairs of planets, "the distance of Mars and Jupiter according to a four-sided figure, the distance of the earth and Mars according to a five-sided figure, the distance of Venus and the earth according to a six-sided figure," and so on.[110] After realizing that he would never obtain "a reason why there are six moving orbs rather than twenty or one-hundred,"[111] however, Kepler turned to what he considered a more exclusive form of explanation. Aware that "only five figures among infinite others could be found for the number and proportion" of the six planetary paths "put forward by Copernicus,"[112] Kepler inserted a regular polyhedron in between each pair of planets. He told his reader how to do this in the following order:

> The earth, as a circle, is the measure of all. Circumscribe it with a dodecahedron. The circle surrounding that will be Mars. Circumscribe Mars with

---

108 *JKGW*, 1, 11.34–35.

109 Ibid., 11.35–12.3: "Igitur die 9. vel 19. Iulii anni 1595. monstraturus auditoribus meis coniunctionum magnarum saltus per octena signa, et quomodo illae pedetentim ex uno trigono transeant in alium, inscripsi multa triangula, vel quasi triangula, eidem circulo, sic ut finis unius esset initium alterius. Igitur quibus punctis latera triangulorum se mutuò secabant, iis minor circellus adumbrabatur. Nam circuli triangulo inscripti radius, est circumscripti radii dimidium. Proportio inter utrumque circulum videbatur ad oculum penè similis illi, quae est inter Saturnum et Iovem."

110 Ibid., 12.4–12.6.

111 Ibid., 12.10–12.

112 Ibid., 13.1–5.

> a tetrahedron. The circle surrounding that will be Jupiter. Circumscribe Jupiter with a cube. The circle surrounding that will be Saturn. Now inscribe an icosahedron inside the earth. The circle inscribed within that will be Venus. Inscribe an octahedron inside Venus. The circle inscribed within that will be Mercury. You now possess the cause of the number of planets.[113]

Although they were conceived metaphysically, Kepler positioned the polyhedra among the planets in order to determine the physical structure of the cosmos with impressive precision. Limited in number to five, they explained why the cosmos contained only six planets and why they were positioned in the way they were around the central body of the sun.

Kepler thus highly prized the regular polyhedra as metaphysical archetypes. In his survey of the heavens, he suggested that the same principles were realized in the form of other celestial phenomena as well. Kepler identified "something resembling a natural faculty" in *On the New Star* as the efficient cause of this realization, "a sort of fervor" that he saw as the source of the new star and other celestial novelties.[114] Accompanied by heat and light, this faculty was known as a natural ability to produce new forms. Kepler noted that it was also identified by natural philosophers in the process of putrefaction, where it brought about "the restoration of an old, rotting form, as if dying, into something new."[115] Kepler accepted a single faculty in the heavens split between the interplanetary ether and the sphere of the fixed stars. "There is in the whole ethereal substance," he wrote, "one faculty where the planets move and another faculty where the fixed stars stand."[116] Since this faculty was similar to "a natural faculty in living beings,"[117] Kepler turned to another anatomical analogy to elucidate it.

Kepler suggested that the natural faculty purified the ether in the same way that a faculty in our eyes ensured the clarity of our vision. In his account of "how the pellucidity of the ether is preserved,"[118] Kepler tied the ability to effect new forms in the heavens to a power that purged the

---

[113] Ibid., 13.18–23: "*Terra est Circulus mensor omnium: Illi circumscribe Dodecaedron: Circulus hoc comprehendens erit Mars. Marti circumscribe Tetraedron: Circulus hoc comprehendens erit Iupiter. Iovi circumscribe Cubum: Circulus hunc comprehendens erit Saturnus. Iam terrae inscribe Icosaedron: Illi inscriptus Circulus erit Venus. Veneri inscribe Octaedron: Illi inscriptus Circulus erit Mercurius.* Habes rationem numeri planetarum."

[114] Ibid., 267.33–268.3.

[115] Ibid., 268.35–37.

[116] Ibid., 269.12–13.

[117] Ibid., 269.13–15.

[118] Ibid., 269.18–19.

ether. Kepler viewed the ether as a very subtle fluid whose fine substance normally offered no observational interference. The heavens, he thought, were kept in their state of transparency by a small degree of heat that also acted in our eyes, facilitating our sense of sight. The natural faculty cleared away areas of condensation in the ether, Kepler wrote, "through the continuation of some very slight heat."[119] Kepler would later give examples of ethereal condensations that had grown to cause observational interference. The sun had been covered by a blood-red cloak for four days in April 1547, he wrote, the same disguise it had worn for a full year following the assassination of Julius Caesar.[120] In either case, Kepler claimed the natural faculty had served to cleanse these condensations. It not only decongested the heavens, however, but also applied the excess ether to the generation of comets and new stars. Essentially, the natural faculty preserved the pellucidity of the ether by collecting "fatty and impure vapors" and turning them into new forms, "a motionless star among the fixed stars" and "a mobile comet among the planets."[121] Since the ether was considered especially concentrated in the Milky Way, Kepler suggested it as a site where the purifying powers of the natural faculty were particularly strong at producing celestial novelties.[122]

Kepler supposed that the soul of the earth was endowed with a natural faculty similar to the one in the celestial ether. In fact, Kepler identified "a certain faculty infused through all of the parts" of the cosmos,[123] a source of continuity that ran deeper than any physical distinction he may have made between the celestial and sublunar spheres. The existence of this faculty in the ether and the ambit of the earth provided an engine for the procreative processes taking place there. It represented another dimension of Kepler's "integrated physics of the heavens and the earth," a conception which has been attributed in part to his "eclectic use of Stoic

119 Ibid., 269.23–27.

120 *JKGW*, 8, 225.13–17.

121 *JKGW*, 1, 269.28–34.

122 Kepler gave two reasons why the Milky Way was not the only site of celestial change. First, he noted that David Fabricius (1564–1617) had discovered a third-magnitude star far away from the Milky Way. Second, Kepler claimed that the frequent birth of comets and new stars in the Milky Way over the course of time would have depleted the ethereal substance so much that the difference would have become apparent. See ibid., 259.19–21: "And so after 1,400 years from the time that Ptolemy described the Milky Way, it seems that something discernible should have disappeared from it, with so many stars having originated from it and been set aflame."

123 *JKGW*, 1, 268.21–23. Here, Kepler is recalling Virgil's "liquid fields" in the *Aeneid*, 6, 726.

ideas."[124] And even if Kepler may not have considered the celestial and sublunar spheres in precisely the same physical terms, the natural faculty served as a source of change that tied the two inseparably together. "What acts in the ether by something nominally hidden," Kepler wrote, "the faculty performs in our own warm air" in a variety of visible things.[125]

Thus, the heavens and earth exemplified the same efficient cause for Kepler through the universal activity of a natural faculty. Did this suggest that the two spheres also shared the same material cause? Perhaps the clearest answer to this question came in the preface to Kepler's *Dioptrice* (1611), a work he wrote in the wake of "the recently invented telescope and the celestial novelties discovered by it."[126] Kepler began the *Dioptrice* with a critical analysis of the prefatory essay by Jean Pena (1528–1558) in the first Latin edition of Euclid's *Optics* (1557). There, Pena had equated the substance of the celestial ether with the sublunar air. Supposing that rays of light descending from the heavens were not refracted from their original direction, Pena concluded that the cosmos was filled with a continuous expanse of air. While Kepler applauded Pena for eliminating any solid celestial sphere that would cause such a change in direction, he accused Pena of "advancing too hastily" and dismissing any physical distinction between the air and the celestial substance "succeeding the air slightly beyond the summits of the mountains."[127] Pena had argued that the air on earth extended to the outer sphere of the fixed stars, contrary to the claim that a series of solid and transparent planetary spheres surrounded the earth "like the white of an egg surrounds the yolk."[128] Although Kepler acknowledged that these spheres were essentially intellectual contrivances, he did not believe this conclusion called for the homogenization of the air and the ether. Atmospheric refraction, first identified by Tycho and especially apparent when objects approached the horizon, was attributed by Kepler to "the upper layer of air" below the moon.[129]

Kepler called on Aristotle as a witness to heat and light in the heavens, employing his account of "a vital heat throughout the universe" in the *Generation of Animals*.[130] Kepler recalled this account in Chapter 24 of

124 Barker, 1991, pp. 138, 154.
125 *JKGW*, 1, 268.23–26.
126 *JKGW*, 4, 334.3–4.
127 Ibid., 335.5–12. On "the air-ether boundary" in the world views of Kepler and Tycho, see Mosley, 2009, p. 145.
128 *JKGW*, 4, 334.27–28.
129 Ibid., 335.13–28. Cf. Bucciantini, 2003, p. 137.
130 *JKGW*, 1, 267.27–29. Cf. Bucciantini, 2003, p. 139.

On the New Star, where he suggested a natural faculty as the source of the new luminary. The new star had "appeared suddenly, diminished gradually, and finally disappeared,"[131] exemplifying the familiar lifespan of a fire. By comparison, Kepler considered a fatty or oily substance, "an excretion" produced by "the globe of the earth," as the ideal fodder for fire.[132] Whether on earth or in the heavens, it appeared that flammable substances were suddenly set aflame. In the light of this similarity, Kepler sought a source of heat "for generating that star in the ether."[133] With the natural faculty in mind, he suggested a spark of life present in all matter, unforeseeably kindled under the right conditions.

Kepler supplied several other examples of terrestrial entities whose spontaneous origins he associated with the new star. As the offshoots of the natural faculty in the earth, Kepler identified metals, minerals, and stones, along with a whole series of small animals. Caterpillars were the product of "the perspiration of trees," moths the manifestation of exfoliated skin, and eels the offspring of swamps.[134] Aqueous animals such as fishes, frogs, and leeches were born from the water, while insects such as bees, flies, and wasps spontaneously emerged from the earth. Underlying all of these offspring was an efficient cause, a natural faculty that seized on superfluous substances and formulated new forms of life. Kepler claimed these substances bore a certain level of moisture that facilitated the natural faculty together with heat. Emitted around the surface of bodies and "occupying their hidden recesses," this moisture fuelled the conception of new forms of life.[135] Kepler pictured this process as the product of "a divine instinct" that saw the natural faculty "order the individual parts towards an end."[136] If it acted this way in the heavens, the natural faculty would account for comets and new stars in the same way that it produced precious stones on earth.

Kepler also compared the heavens and earth to the human body. He held the human body as a well known site of spontaneous generation that turned "the sweat of the head into lice and the sweat of the body into fleas."[137] And just as the male body swelled with seminal fluid with the

131 *JKGW*, 1, 267.32.
132 Ibid., 267.35–268.1.
133 Ibid., 268.2.
134 Ibid., 268.29–31.
135 Ibid., 269.1–4.
136 Ibid., 269.6–7.
137 Ibid., 269.28–29.

arrival of spring, the abundance of the earth during the same season led to the decline of "disease, famine, and the gloom of winter."[138] Every specimen of spontaneous generation provided Kepler with another sign of the vital essence of the earth. By analogy, he suggested that this essence also extended to the celestial ether, now the site of change and growth. The new star seemed to be simply another example of this essence. Kepler made it clear, however, that he thought the natural faculty worked greater wonders in the human body and "the infinite arrangement of parts of living beings" on earth than in the heavens.[139] This was one way he claimed we made up for our meager size in comparison to the magnitude of the heavens. "Far more noble," Kepler wrote, was the occupation of the natural faculty in the human body than the power of the heavens "to roll up celestial refuse into the single form of a new star."[140]

Despite this difference in activity, the diffusion of the natural faculty "from the earth to the heavens" was one way that Kepler approached the celestial and sublunar spheres according to similar parameters.[141] Kepler claimed that a complete absence of parallax, along with the splendor and scintillation of the new star, determined the location of the luminary in the sphere of the fixed stars. While he compared it to an earthly conflagration, Kepler strongly opposed the view of Francesco Patrizi (1529–1597), who suggested that a new star was "a living flame lacking any corporeal substance."[142] Kepler reckoned that the star stemmed rather from a sudden alteration of the ether in a particular part of the Milky Way.[143] Different though the earth and ether were, Kepler attributed the spontaneous generation of new forms to a natural faculty acting everywhere. Condensations of the ether, the air, and the earth—even concentrations of sweat on the human head—furnished the natural faculty with a means of manifesting new forms. The natural faculty thus served as a spark igniting the generative process. While the new star initially surpassed the brilliance of even the brightest stars, the enduring impression of light left by the natural faculty meant that even minerals, stones, and other subterranean substances harbored a lasting source of luminescence. Kepler would later describe how this occurred in the case of precious stones produced

[138] Ibid., 318.3–6.
[139] Ibid., 288.24–28.
[140] Ibid.
[141] Ibid., 288.28–30.
[142] Ibid., 246.25–30. See Boner, 2009b.
[143] *JKGW*, 1, 258.38–259.5.

"among the secrets of chymists." In what he called an "extraordinary and highly memorable experience," Kepler told the story of stones whose luster after exposure "only to the light of day" would last even after being removed to a place of darkness:

> Thus, among the secrets of chymists is the extraordinary and highly memorable experience that occurs when they produce gems ... that lack any light as long as they lie hidden in the dark. Yet if someone exposes them only to the light of day, they are kindled by it like candles and carry a splendor with them even into darkness, shining like the eyes of cats until [their splendor] is extinguished again after a short time.[144]

Shining in the darkness, these stones suggested that the natural faculty instilled an affinity to light in substances well below the surface of the earth.

The universal nature of the natural faculty expressed a fundamental unity in Kepler's world view. The acceptance of unforeseen events in the heavens did not deter Kepler from discerning a comprehensive order in the cosmos. In fact, the new star of 1604 gave him even greater reason "to apprehend the assembly of connections unifying the heavens and earth" on his way "to discovering the plan of creation."[145] The ether was enormously more subtle than any earthly substance, Kepler argued, yet the universal presence of a natural faculty allowed him to assign the new star to the same spontaneous procedure taking place on earth. Although the new star had appeared suddenly and seemingly out of nowhere, Kepler claimed it came from an ethereal condensation of the Milky Way that fulfilled the same function in the heavens that he found on earth in the continual cleansing of matter. Spontaneous generation was a process that saw the natural faculty realize new forms according to the same underlying principles. Kepler employed this ontological continuity in his physical inquiry of the heavens, despite any material differences he may have identified between the celestial and sublunar spheres.

---

144 *JKGW*, 6, 273.28–33: "Sic inter Chymistarum arcana est mirabile hoc et imprimis commemorabile experimentum, quod gemmas ... apparant, quae cùm lateant in tenebris, ut alia lumine cassa; si quis tamen illas luci solius diei exponat, incenduntur eâ ut candelae, splendoremque secum in tenebras etiam deferunt, lucentes ut oculi felium; qui brevi tempore iterum extinguitur."

145 Simon, 1979, p. 9.

Philosophical Marvel and Theological Miracle: The Many Meanings of the New Star

Kepler opposed the prevalent theological view that the new star had been "a miracle created by God from nothing."[146] He argued that the star had always possessed a material substance whose change in form had stemmed from the natural faculty among the fixed stars. Kepler accepted that creation from nothing was "the work of God alone,"[147] yet the material nature of the new star did not disqualify it from being the direct consequence of divine intervention. That the appearance of the new luminary was merely the product of material coincidence Kepler condemned as "an affront to God Almighty."[148] While he studied the physical features of the new star scrupulously, Kepler recognized that "there still remained a place for God to arrange the circumstances of place and time."[149] It was anything but accidental, he argued, that the star had appeared so close to the three superior planets from our particular perspective on earth. Coincidence so extraordinary could not be taken lightly. And even if the new star was shown to bear little significance for us, there was still no mistaking the author. Kepler wrote that the human race had been rightfully moved by the new luminary as a sign of God, "the author of the world" who spoke through the marvels of nature:

> Thus, he who acknowledges God as the author of the world when he attributes to Him the order of all of nature and every one of her individual works, including of course this star and the admiration of men that accompanies it, will certainly not fear that God, established in this way as the author of that marvel, should be angered by such an admiration of men.[150]

Kepler did not assign any specific meaning to the new star. He considered a number of possibilities, including the conversion of the new world to Christianity, the ruin of the Islamic religion, the coming of the Messiah of the Jews, and the advent of Christ. In the end, however, he did not endorse any one prediction in particular. Kepler considered this task a

146 *JKGW*, 1, 267.10–12.
147 Ibid., 267.14–15.
148 Ibid., 284.18–21.
149 Ibid., 284.16–18.
150 Ibid., 283.36–40: "Qui ergo Deum Mundi authorem fatetur, cùm et totius naturae ordinem, et omnia eius opera singularia, nimirum hanc quoque stellam, et quae hanc consequitur, admirationem hominum, eidem Deo transcribat: equidem hoc non metuet, ut Deus ille, huius prodigii hoc modo constitutus author, irascatur huic hominum admirationi."

conjectural exercise and ultimately a lesson in futility. “If it had pleased God to signify openly to men what He willed,” Kepler wrote, “He would have inscribed it in the heavens with written words. And so men struggle in vain to decipher the divine will.”[151] With no specific meaning in sight, Kepler suggested the new star should recall individuals to their own lives rather than the risk of revolution. He perceived the star as a divine puzzle, inviting “a closer inspection of our own activities” and encouraging us to search for something hidden.[152] Kepler considered any prediction of political circumstances, “in the light of the magnitude of our miracle,” “clearly worthless.”[153] In support of this view, Kepler referred to Helisaeus Roeslin, who had apparently found no political event that measured up to the magnitude of the new star of 1572. “Roeslin recognizes that nothing that has happened since 1572 until now,” Kepler wrote, “may be rightly compared to the magnitude of that miracle.”[154] Tycho had also claimed that no political upheaval, not even the uprisings in Belgium, Britain, and Hungary, had equaled the enormity of the new star of 1572. Now in the case of the new star of 1604, Kepler held out little hope of something so extraordinary in the political sphere that could match the celestial marvel. Kepler could not completely deny such a possibility in the greater sphere of humanity, however. The birth of Jesus had occurred under the sign of a new star, after all, which had also marked the onset of a Fiery Trigon 1,600 years earlier. In the light of this link, Kepler called on others to join him in sifting through “the oracles and prophecies of peoples” to see if there could be a connection so great as the birth of Christ:

> Come, let us examine with the astrologers at length the greatest things in the hands of the human race. Let us search the oracles and prophecies of peoples to see whether there may be one of them now at hand that is divinely signified by that splendid star. Let us imitate the example of the Magi, who had in their possession the prophecies of Daniel of the coming of the Messiah when an extraordinary star appeared around the time of the great conjunction, at the beginning of the sixth period of triplicities, and were stirred to study books and … seek Him who was born under the sign of a star, King of the Jews, Savior of the world.[155]

---

151 Ibid., 346.38–39.

152 Ibid., 347.1–4.

153 Ibid., 347.5–7.

154 Ibid., 347.9–11. On Kepler's running debate with Roeslin over the significance of the new star, see Granada, 2005 and 2011b.

155 *JKGW*, 1, 347.14–22: “Age itaque perquiramus cum Astrologis vel tandem, quae sunt penes humanum genus maxima: excutiamus gentium oracula, prophetiasque, utrum aliquid eorum sit, quod nobis iamiam instans per hoc tam splendidum sidus divinitus sig-

David Fabricius (1564–1617), who determined the influence of the new star according to the ascending sign of Aries, criticized Kepler for his predictive imprecision.[156] Confident that his own predictions would be proven true, Fabricius asked Kepler in a letter of 1607 to attend more closely to "that noblest part of astrology" that would link particular things to the new luminary.[157] This part dealt with "the configuration of the heavens, the houses, the aspects, and other similar things," Fabricius wrote.[158] Although he admitted that astrology was "polluted by many fictions," Fabricius was not discouraged from determining the influence of the heavens "so that the truth be revealed."[159] He noted that when the new star first appeared, Aries, "the sign of Germany," was ascending, while the sun was setting.[160] Fabricius suggested that this configuration signified "peace and a miraculous mutation of the empire for the better."[161] For his own part, Kepler saw the association of Aries with the eagle of the German Empire, whose emblem had served as Fabricius's source of interpretation, as "excessively subtle."[162] Kepler suggested several signs whose celestial positions could just as well have signified the same thing, "such as Capricorn, since it was in the midheaven, Pisces, since it was ascending when the new star was culminating, and Gemini, since it was ascending when the new star was descending.[163] Kepler claimed that Fabricius had not put forward a prediction but merely expressed under their guise "grievances about his neighbors, his opinion concerning the condition of the empire, and his desire for retribution and repair."[164] Kepler further accused Fabri-cius of fulfilling his civil duty by simply attempting "to scare away insurgents and bad neighbors on the one hand while encouraging the Emperor on

nificetur. Imitemur exemplum Magorum, qui cum haberent oracula Danielis de venturo Messiah, orta stella mirabili, eaque ipsa quoque sub magnae conjunctionis tempus, quando cepit sexta Periodus triplicitatum; excitati sunt ad inspiciendos libros … ad quaerendum, quem, stella Indice, natum putabant, Judaeorum Regem, mundi Salvatorem."

156 Kepler considered Fabricius Europe's finest observational astronomer after the death of Tycho. See ibid., 210.33–36; cf. Christianson, 2000, pp. 273–276. On Kepler's conversation with Fabricius concerning the generation and meaning of the new star, see Granada, 2011a.

157 *JKGW*, 15, no. 430, 20–24.

158 Ibid., 20–23

159 Ibid.

160 Ibid., 301–302.

161 Ibid., 300–307. Cf. Granada, 2011b, p. 74.

162 *JKGW*, 1, 342.21–23.

163 Ibid., 342.23–26.

164 Ibid., 354.25–28. Cf. Granada, 2011b, p. 80.

the other."[165] In response, Fabricius requested that Kepler weigh matters more carefully: the sign of Aries was a clear indication of the condition of the empire, whose well being he valued far more than any personal or provincial concern.

Rather than proffer any alternative prediction, Kepler emphasized that the extraordinary appearance of the new star had been "afforded by the finger of God in the highest heaven."[166] Kepler wrote that the new luminary had "emerged like a ray of light of remarkable omnipotence" that had "employed the service of nature."[167] It was not contradictory, Kepler claimed, to accept the natural conception of the new star as divinely inspired. All Kepler had to do was turn to the Book of Genesis, where he found "God telling nature, 'Produce,'" and nature, "previously dead to producing," took up a new faculty for fulfilling that very task.[168] By analogy, Kepler concluded that a natural faculty in the heavens had actualized this divine decree in the form of the new star. What the spatiotemporal circumstances of it suggested for the future, however, Kepler could not say. He limited his role to the wonders of philosophical worship, awaiting the word of God before moving forward with a more precise prediction. "If divine law is to be implored," Kepler wrote, "I pray that God guide me to explain to men what He wills with this star."[169]

While Kepler interpreted the natural conception of the new star in terms of secondary causes, he considered these causes the devices of the divine will. Kepler claimed it was by divine decree, the deliberation of "God the architect," that the new star had come "into the company of the planets" according to our perspective on earth.[170] "For who shall deny that the governance of everything that employs the service of nature and was created by it," Kepler asked, "is within the power of God?"[171] Convinced that the natural causes of the new star were used to express a providential message, Kepler stopped short of suggesting any specific meaning. Instead, he argued that something so extraordinary served as an ideal opportunity for the human race to reflect on their spiritual condition. Apart from this, there was no doubt that divine providence continued to play a part in the repentance of sins:

---

165 *JKGW*, 1, 354.28–31.
166 Ibid., 291.34–35.
167 Ibid., 291.27–29.
168 Ibid., 291.30–32.
169 Ibid., 291.32–34.
170 Ibid., 290.1–3.
171 Ibid., 290.5–6.

> With infinite things occurring everyday, each according to its own natural cause, does not a place still remain for divine providence, if in other things, then especially in administering the retribution of sins? In this way, what happens not only follows the preceding causes in nature but also divine justice.[172]

Kepler did not suggest any further significance for the new star beyond the fact that it served as a sign of our own fragility. Any other interpretation of the star he regarded as deceitful. "Let God alone be true," Kepler wrote, "while every man—and so also me and this interpretation of mine of that prodigy that I ponder—a liar."[173] While the meaning of the new luminary may have escaped him, Kepler claimed there was no questioning the author. In his survey of other opinions near the end of *On the New Star*, Kepler looked favorably on the theological view that the new luminary had been the direct result of divine providence:

> The Son of God took on our entire nature, including every property except for sin, along with the full series of interior faculties of the human soul, where speech, various languages, and all of the arts and human disciplines find their origin, in the womb of the blessed Virgin, in the unity of the person. And so we may believe that this same Lord and God of ours, whose delight was so great that His commerce with the sons of men shall be eternal, as even today He certainly does not cease from public signification of His concern for us, arranged and ordered this signification, expressed in the new star according to an account of place and time, in such a way that it could neither remain hidden from us nor fail to have an extraordinary effect on us, especially scholars and astrologers.[174]

---

172 Ibid., 290.6–10: "Nonne infinitis rebus quotidie evenientibus, quâlibet ex sua causa naturali, locus tamen providentiae divinae relinquitur, cùm in aliis, tum praecipuè in exercenda vindicta scelerum; quâ efficitur, ut id cadat, quòd non solùm antegressis in natura causis, sed etiam justitiae divinae sit consentaneum."

173 Ibid., 292.1–3. Cf. Granada, 2011b, p. 79.

174 *JKGW*, 1, 280.13–24: "... quòd Dei Filius ipsam hanc naturam nostram totam et integram cum omnibus proprietatibus suis, dempto peccato, cumque totâ interiorum facultatum animae humanae serie, unde sermo, variae linguae, omnesque artes et disciplinae humanae profluunt, ex utero beatae Virginis in unitatem personae suscepit. Credibile est igitur, eundem illum Dominum et Deum nostrum, cuius tanta fuit delectatio, tantum in aeternum erit commercium cum filiis hominum, etiamnum hodie non planè cessare à publica significatione suae de nobis curae, eamque significationem, in nova stella propositam, sic ordinasse et instruxisse, per descriptionem temporis et loci; ut non posset nos, praesertim literatos et Astrologos ... vel latêre, vel non summoperè commovere."

CHAPTER FOUR

THE COMETS OF 1607 AND 1618

In the serene sky above Prague shortly after a fireworks display on 16/26 September 1607, Kepler first observed another celestial novelty. "Without spectacles," Kepler wrote, it "appeared to give off a light equal to the fixed stars."[1] He duly noted the location of the new luminary and the fact that his fellow observers identified a tail. Kepler would later explain the disappearance of the tail as a product of the earth's point of view, as the line of observation drew parallel with the line along which the tail of the comet extended. The swift motion of the comet slowed steadily in early October and then sped up, accelerating to an eventual disappearance at the end of the month.

Kepler quickly published a report in German on the nature, origin, and astrological significance of the comet. Convinced that it could cause changes in the normal course of sublunar nature, Kepler suggested "a sympathy with the heavens" as the source of the earth's response.[2] In some cases, the appearance of a comet could move the earth "to perspire many moist vapors," resulting in "heavy rain and flooding."[3] Kepler claimed the effect this had on the air could lead to a number of health problems, including "headaches, dizziness, catarrh, as in the year 1582, and even pestilence, as in the year 1596."[4] In other cases, earthquakes could result from an abundance of the hot and dry exhalation violently forcing a way out of the terrestrial interior. Kepler confined his conjectures to what he considered the natural causes of the comet, whereby the earth and "all of nature" sensed "something extraordinary in the heavens" and was "frightened" into a form of heightened activity.[5] Kepler would later combine his claim to natural causes with a criticism of other astrologers who did not act similarly. He complained that the principles they deployed to determine the influence of the comet fell fully outside the course of nature. What they claimed did not appear to concern "natural causes," Kepler wrote,

---

1 *JKGW*, 8, 157.3–8.
2 *JKGW*, 4, 62.19.
3 Ibid., 19–22.
4 Ibid., 22–24.
5 Ibid., 16–18.

but rather "an entirely different form of prognostication" that "made comets simply the signs of future things."[6] Although many astrologers considered such characteristics as the color, path, and position of the comet, Kepler argued that these characteristics were interpreted as if the comet were aware in some way of their impact, an assumption that could not be further from the truth. "A comet traverses the ether in a straight line," Kepler wrote, "ignorant (if you now pretend that it is a man making use of reason) of what its motion shall look like here on earth."[7]

That comets were almost always "far above the Moon and deep in the heavens" Kepler accepted as "clearly proven by Tycho Brahe."[8] In his account of the comet of 1577, Tycho had argued against "the majority of the learned" by claiming that comets were not meteorological phenomena made up of dry and heavy earth drawn up and set aflame "in the warmth of the swiftness of the air."[9] Tycho criticized advocates of this view for simply acquiescing to "the authority of Aristotle" without making "any observation with sophisticated instruments or mathematical demonstration" themselves.[10] Intent on investigating their location through observations that he "submitted to arithmetical calculation," Tycho concluded that "comets were shown by certain reasons to be in the ether."[11] For evidence, Tycho turned to the comet of 1585, whose absence of parallax suggested a location well beyond the sublunar sphere. Tycho explained that the comet had displayed "scarcely a single minute of parallax" and that Christoph Rothmann, a mathematician at the court of Wilhelm IV (1532–1592), Landgrave of Hesse-Kassel, had come to the same conclusion.[12] Tycho wrote highly of Rothmann as "a man exceptionally skilled in astronomy" who agreed that comets existed exclusively in the celestial ether.[13] The interplanetary course of comets had also convinced Tycho to reject the existence of the celestial spheres and accept that "the heavens were not made up of hard and impenetrable matter."[14] Tycho even considered the

6 Ibid., 21–25.
7 *JKGW*, 8, 231.34–36.
8 *JKGW*, 4, 59.19–20.
9 *TBOO*, 4, 7.2–13.
10 Ibid., 7.8–13.
11 Ibid., 7.14–32.
12 Ibid., 223.29–35.
13 Ibid.
14 Ibid., 223.39–42. On the comet of 1585 and how it contributed to Tycho's rejection of the solid and impenetrable celestial spheres, see Granada, 2006.

possibility of a "science [*scientia*] of regular motion divinely introduced" in comets and "perpetually preserved" by their paths.[15]

Kepler dedicated the second part of his report on the comet of 1607 to its appearance, path, and possible influence. He noted that a "faint pale glow" had led some to compare it to the moon.[16] This was a feature shared by countless other comets, which Kepler saw as the source of why so many astrologers thought comets were the cause of heavy precipitation and sickness. Others claimed that the comet had first appeared in the fiery sign of Leo and disappeared in Sagittarius, another fiery sign. This had led astrologers to predict barren and dry weather along with war, contrary to the simple fact that the comet had first been seen "without a doubt in the sign of Cancer."[17] In the end, Kepler disapproved of any association with the signs of the zodiac and delicately referred his readers to a place where they would find his full critique. "What we should make of this division of the signs," Kepler wrote, "may be found in my book, *On the New Star*."[18] Kepler would be more direct and dismissive when it came to determining the situation of the comet in the system of houses. This question involved horary astrology, a form of "fortune telling" that Kepler found "not only wrong, but even childish and vain."[19] Kepler gave far greater priority to the actual path of the comet (apart from the motion of the earth) and "whether it existed above or below the moon."[20] He claimed that these questions were not meant "for the German reader," however, but belonged to a book that he would later publish in Latin.[21]

For the purposes of his report, Kepler simply set out to show that comets possessed their own form of motion. They could not remain motionless on the plane of the ecliptic, he claimed, since they did not trace a large circular path among the planets, nor could they remain motionless outside the plane of the ecliptic since they did not trace a small circle elsewhere. Kepler imagined a countless multitude of comets moving day and night "through the great amplitude of the heavens" with only a few ever coming into view.[22] He maintained that their motion was linear and that the apparent curvature of their course was an illusion produced by

15 *TBOO*, 4, 223.42–224.3.
16 *JKGW*, 4, 74.29–30.
17 Ibid., 35–36.
18 Ibid., 38–40.
19 Ibid., 74.41–75.1.
20 Ibid., 75.6.
21 Ibid., 75.5–8.
22 Ibid., 65.32–33.

the diurnal displacement of the earth. Kepler would later write in Latin that every comet followed "a trajectory on a straight line continuously forward," distorted by the annual motion of the earth "that Aristarchus and Copernicus attributed to it."[23] Kepler did not adopt Tycho's idea of a divine "science [*scientia*]" to explain this linear motion. On the contrary, he compared the course of every comet that approached the earth to the path of a beached whale. "Without knowing the land it encountered or the message it brought the people there," a beached whale blindly worked at breaking free and "returning to its element."[24] How it made its way there in the first place, Kepler thought, pointed to the existence of "an invisible spiritual creature that led it there."[25]

Speculation about motive spirits provoked the censorship of the theological faculty at Leipzig, where Kepler planned to publish an enlarged Latin edition of his report. Sent to Leipzig for lack of suitable printing facilities in Prague, the new edition would remain among Kepler's unpublished papers, "awaiting another occasion."[26] In particular, the theological faculty condemned the suggestion of something that could pass in and out of existence after the original act of creation. Joachim Tanckius (1557–1609), a professor of anatomy and surgery whom Kepler appointed to administer publication of the manuscript, informed Kepler that "our theologians allow neither the generation of new creatures after the original act of creation nor their expiration."[27] Georg Weinrich (1554–1617), the Dean of Theology, expressed his concern to Tanckius over the inference that spirits would "multiply in the same way as men." In an age that witnessed an unusually high number of comets, Weinrich may have felt additional urgency to question an already controversial claim. As Tanckius reported to Kepler, Weinrich did not reject the active presence of spirits in comets but rather their creation and permanent passing:

> Weinrich adduced from Sacred Scripture that God does not create new spirits. In the beginning, He created good spirits and some defected from the good while others remained that way. Good spirits serve man while the evil ones cause him injury. The same would follow if you suppose that new [spirits] are not created by God but permitted by Him to provide guidance. Spirits do not multiply in the same way as men. And as for the expiration of

---

[23] *JKGW*, 8, 142.22–25.

[24] *JKGW*, 4, 65.8–10.

[25] Ibid., 65.11–12. On Kepler's account of cometary spirits, see Hübner, 1975, pp. 18–19; Yeomans, 1991, pp. 52–53; Boner, 2012b.

[26] *JKGW*, 8, 133.30–36.

[27] *JKGW*, 16, no. 479, 16–18.

> those spirits, if they are not created they do not expire but endure for other purposes when the comets expire.[28]

Convinced that new spirits were not created by God, Weinrich still remained uncertain whether comets were "governed by spirits or conveyed by a natural motion."[29]

In addition to the censorship it suffered, the enlarged Latin edition experienced delays that left the printer "no hope of profit."[30] With interest in the comet waning and the death of Tanckius in November 1609, publication was finally abandoned. In a poem that Tanckius penned for the preface to the new edition, he praised Kepler for "revealing fates, marvels, and the very miracle of nature to the impure world."[31] Tanckius denounced those who "sullied the mind with the blemish of an age-old stain" and commended Kepler for improving on the insight of earlier astronomers.[32] In a letter to Kepler, Tanckius had already expressed his loyalty to a new view of comets that broke with the old theory of Aristotle. "They are immersed in the mire of exhalations," Tanckius wrote of his Aristotelian colleagues, and "prefer to err with Aristotle... Let us leave them to fumble with the husks while we seize on the kernel."[33] Later in his poem, Tanckius would elevate Kepler high above "in the light of the heavens," where he stood "scrutinizing the order of the spheres."[34] This poem would remain unpublished until the appearance of Kepler's *On Comets* (1619–1620), a compound study of the comets of 1607 and 1618 that featured an enlarged edition of his original report.

The Role of Divine Providence in Kepler's Cometary Theory

Kepler continued to honor the wishes of Rudolph after the abdication of the Emperor in May 1611. As Kepler later reported in a letter to Peter

---

[28] Ibid., 27–33: "Adduxit ex Sacris Deum non creare novos Spiritus: Spiritus primo creavit bonos, ex bonis quidam defecerunt, quidam in statu permanserunt. Boni hominibus serviunt: Mali laedunt. Si posuisses, non creari à Deo nouos sed permitti à Deo eos regi, ferre id potuisset: Nec spiritus eo modo ut homines sese multiplicant. Deinde; Spiritus illos evanescere: Si non creantur, non evanescunt, sed evanescente Cometa permanent ad usus alios."

[29] Ibid., 33–34.

[30] *JKGW*, 8, 133.35–36.

[31] Ibid., 136.16–18.

[32] Ibid., 136.5–21.

[33] *JKGW*, 16, no. 472, 7–9.

[34] *JKGW*, 8, 136.15–23.

Crüger (1580–1639), Rudolph was "unwilling to consent to his departure from court," so Kepler stayed on, clinging to "the empty hope" of a position in the Duchy of Saxony.[35] When Rudolph died in January 1612, his brother and successor, Matthias (1557–1619), immediately re-appointed Kepler as Imperial Mathematician. With his "money consumed" and his mind "forgetful of astronomy,"[36] Kepler was finally granted permission to leave for Linz, where he took up the position of District Mathematician. Kepler departed Prague in April 1612 and arrived in Linz a month later.

Kepler met with auspicious astrological circumstances soon after his arrival in Linz. He would later note in a letter to a noble that in the fall of 1613 Saturn and Mars were in positions of power. Configured in such a way, they recalled the form of the heavens at Kepler's moment of birth. Kepler thought this would result in the accentuation of tendencies already present in his personal nature. "By the cause of the stars," he wrote, his early years in Linz could be marked by a greater display of familiar qualities, such as his "show of piety and compassion, the pursuit of fame through new and unusual ideas," and his "spiritual anxiety over salvation."[37] Considering the promise of this prediction, Kepler may have been elaborating on those qualities that made a noble and potential patron more aware of his predicament. Even in his account of his second marriage in October 1613, Kepler complained of his complete commitment to study as a frustrating prospect for his future wife. "In my person," Kepler wrote, "is a man scarcely political, solely dedicated to study. There was neither security nor any hope of settling things."[38]

The conviction that the configuration of the heavens could intensify a particular proclivity did not amount to a fundamental change in one's character. "Let me be an example," Kepler wrote, "that God rarely alters the course of nature."[39] Divine providence was expressed in the orderly course of the stars, he argued, which determined the sequence of days and succession of seasons without deciding the details of an individual life. Kepler considered the regular motion of the heavens as the focal point of mathematics and natural philosophy. Astrology, on the other hand, was the study of how the sublunar sphere reacted to the heavens by intensifying certain activities and attributes. Aside from this amplification, the

---

35 *JKGW*, 17, no. 710, 28–30.
36 Ibid., 30–35.
37 *JKGW*, 17, no. 669, 83–88.
38 Ibid., 49–50.
39 Ibid., 88–89.

earth and many other living beings occupying the sublunar sphere were fully responsible for their responses. This was particularly important when it came to evil, Kepler explained, since evil was the creation of the human will alone. "Everything God created is good," Kepler wrote, and "neither the heavens nor their light rays are to blame, but rather original sin and the evil will."[40] Kepler claimed that the configuration of the heavens at his birth had produced particular tendencies, with a sextile of Saturn and the Sun coloring a considerable part of his temperament.[41] He believed that the periodic return of a celestial body to a birth position reinforced the original impression. For Kepler, reinforcement came in the return of Saturn in the fall of 1613 to "the highest part of the zodiac,"[42] where it foretold the enhancement of his personal features. This idea of astrological influence remained in keeping with the principle of general providence, which Charlotte Methuen has distinguished from the principle of "special providence" in her study of sixteenth-century Lutheran theology.[43]

The three comets that appeared in 1618 came in quick succession. As soon as he learned of the first comet in August, Kepler desperately scanned the sky above Linz. Despite his poor vision and the interference caused by the hot summer air, he finally "chanced on the form of a comet that had already lost almost its entire tail."[44] Kepler recalled the obscurity and eventual disappearance of this comet in September, scarcely perceptible by even an attentive astronomer, "let alone the general public."[45] The second comet that appeared in November was often confused with the third one, which came into view around the same time and place. "I see that those two [comets] are confused as one," Kepler wrote, "since they appeared at the same time and advanced from the same part of the heavens."[46] Kepler noted, however, that while the second comet did not endure "beyond the end of November," the third one did not disappear until after "a good part of January."[47] Particularly in the case of the second and third comets, Kepler marshaled a mass of observations he had received as part of a network of astronomical exchange that extended to Ingolstadt, Innsbruck, Magdeburg, Strasbourg, Wittenberg, and beyond.

40 *JKGW*, 4, 225.24–30; cf. Rabin, 1987, p. 181.
41 See Kepler's letter to Herwart von Hohenburg, *JKGW*, 13, no. 117, 253–260.
42 *JKGW*, 17, no. 669, 79–80.
43 Methuen, 1999.
44 *JKGW*, 8, 177.25–26.
45 Ibid., 177.7–8.
46 Ibid., 178.27–28.
47 Ibid., 177.9–11.

Kepler claimed that these comets had come as a call to reflect on the spiritual condition of the human race. "God displayed the same sign in the heavens to everyone," Kepler wrote in his study of the comets of 1618, either to admonish individuals "or to announce a great universal calamity" whereby those same individuals "might be set straight."[48] We may define this as the spiritual frontier of what Gérard Simon has described as Kepler's search for "rationalizing divination."[49] The many signs of divine providence that Kepler identified in the celestial sphere did not signify any future event explicitly. "Let him speak who knows the spirit of God," Kepler wrote. "The only thing I know for certain is the weakness of my conjectures."[50] Rather than proffer a clear prediction, Kepler accepted a complementarity between divine decree and natural causes.[51] He argued that comets became visible to the earth without any "distortion in the laws of nature,"[52] and that the rarity of their appearance reminded everyone of their religious calling.

Kepler echoed the words of his theology professor from the University of Tübingen, Jacob Heerbrand (1521–1600), who declared that divine providence involved "a direct application of natural causes."[53] Although Heerbrand understood the comet of 1577 as an arbitrary and unpredictable act of divine intervention, he accepted it as part of the natural world and the product of secondary causes. Seen in this way, the exclusion of the comet of 1577 from the regular course of the cosmos did not deter Heerbrand from the opinion that "the universe could be regarded as revelatory."[54] Although Kepler supposed that the celestial ether was populated by a countless number of comets, he suggested that the circumstances surrounding their appearance were divinely inspired. And in his account of what he called "the natural significance" of the comets of 1607 and 1618,[55] Kepler assumed that comets, as part of a larger providential plan, were still subject to natural philosophical enquiry. Their daily displacement and plane of trajectory were, among other things, the subject

48 Ibid., 261.7–9.

49 Simon, 1979, pp. 67–68.

50 *JKGW*, 8, 258.13–14.

51 Simon has suggested a "place between material causality and miraculous intervention," which in the case of comets Kepler filled with a spiritual principle. See Simon, 1979, p. 67.

52 Ibid.

53 Methuen, 1999, p. 104.

54 Ibid., p. 107.

55 *JKGW*, 8, 238.4.

of scrupulous study while at the same time signs of God's greater plan. Kepler claimed that comets and new stars, as a "class of portents," were not contrary to, "but rather beyond the ordinary nature of the heavens."[56] While the rarity of their appearance reflected a preternatural coincidence of time and place, Kepler found "evident causes" for their rectilinear motion."[57] To explain how these causes led to a natural course of motion that still bore the mark of providence, Kepler returned to his beached whale analogy. Like a whale that accidentally arrived "on dry land, struggling to breathe," a comet that came near the earth followed a course as blindly as any other. By this analogy, Kepler also compared the origin of comets in the ether to the spontaneous generation of whales and other "sea marvels" in the ocean:

> A whale is a part of nature insofar as it is a kind of animal whose life and origin is in the immense expanse of the ocean. In fact, so many new sea marvels are found so frequently that it cannot be the case that their species were all created at the same time in the beginning and propagated only through the intercourse of male and female down to this day... Some correctly call the ocean 'the father of sea marvels,' for it was originally created for the purpose of producing so many and various marvels. Be that as it may, when such a sea marvel strays and is washed up on shore, left struggling to breathe on dry land, the matter is accepted as an omen. It is then agreed that such things follow that bring verisimilitude to that sign. None of these things are known to the fish, however, neither the place where it will arrive nor what it will announce there. All that the fish knows is that it wants to escape and return to the water, free from the sand, to save its own life.[58]

While Tycho had also attributed comets to divine providence, he claimed that they did not derive entirely from natural causes. In his analysis of the comet of 1577, Tycho denied that comets originated from "a natural order," whereby they emerged as "strictly natural products" in a celestial

---

[56] *JKGW*, 1, 341.1–3.

[57] *JKGW*, 8, 216.37–39.

[58] Ibid., 232.32–233.5: "Cete naturae pars est, genus quippe animalium, cuius vita in immensa Oceani amplitudine est, ortus verò indidem. Tot enim quotidie nova visuntur monstra ut non sit verisimile, omnes species eorum initiò simul creatas, solaque consuetudine maris et femellae hucusque propagatas... Itaque quidam Oceanum rectè monstrorum patrem appellant: In hunc enim finem initiò conditus est, ut ista varia et quotidiè nova monstra produceret. Haec etsi ita habent, tamen si quando monstrosa huiusmodi moles aberrans littoribus infunditur, et reciprocante aestu destituitur in sicco: res in omen accipitur, et constat, talia sequi, quae significationi illi verisimilitudinem concilient. Atque nihil horum constat ipsi pisci, nec in quam regionem, nec quid nunciatum advenerit; quin potius quaerit, qua evadat, liberatusque arenis humido elemento potiatur, ut vitam tueatur suam." Cf. *JKGW*, 4, 64.40–65.11.

ether subject to change.[59] Although he accepted generation and corruption in the heavens, Tycho defined comets as "a divine transgression of the natural order through a direct and immediate intervention of God."[60] He conceived of comets as "a special creation by God," "a divine marvel" and "miracle in the nature of the heavens" that obliged us to admit our ignorance when it came to their origin.[61] Tycho claimed that the material conception and constitution of comets were matters that we humans, "with our limited earthly understanding,"[62] could never fully grasp. By defining comets in this way, Tycho appears to have gone about their physical study with greater caution than Kepler.[63]

Matthias Hafenreffer (1561–1619),[64] another theologian who taught Kepler at the University of Tübingen, also argued that divine intervention in the order of nature was ultimately incomprehensible. According to his view of divine providence, Hafenreffer claimed that comets and other extraordinary portents could not conflict with any established system of physical causes. Rather, comets were an exception to the rule when it came to the causal explanation of events.[65]

Although he pinned down the principle of divine providence to natural causes, Kepler agreed with Hafenreffer that it was an unpredictable part of God's plan. Kepler had said the same in *On the New Star*, where the circumstances surrounding the new luminary could not completely account for the level of coincidence involved in the proximity of the superior planets—divine providence had also played an important part. Near the end of *On the New Star*, Kepler finally reached a point where nature gave way to God, whom he revered as "the author of nature" and the source of "the circumstances of place and time." This reverence accompanied the suggestion that the new star had not simply been "a roll of the dice,"[66]

---

[59] Granada, 1997, p. 396; cf. Christianson, 1979.

[60] Granada, 1997, p. 396.

[61] Ibid., pp. 396–397.

[62] Ibid., p. 397; cf. Segonds, 1993, p. 373.

[63] Cf. Methuen, 1999, p. 104.

[64] Kepler remained in contact with Hafenreffer until the latter's death in October 1619. In one letter, Hafenreffer praised his former pupil's "eminent and noble knowledge," which extended to "every lofty thing, all the way up to the outer surface of Saturn." When it came to things that were "celestial in the spiritual sense," however, Hafenreffer counselled Kepler that the human intellect would fall forever short. "As for those things that are, in a word, theological," Hafenreffer wrote, "enough! Here, every ingenuity of the human mind must be foolish." See *JKGW*, 17, no. 829, 14–19; cf. Hübner, 1975, p. 282.

[65] Cf. Methuen, 1999. On Hafenreffer's distinction between astronomy and natural philosophy as two separate disciplines, see Rothman, 2011, p. 117.

[66] *JKGW*, 1, 284.7–8.

but a divine miracle that God had revealed to the human race as a form of admonition:

> For although I dedicated several earlier chapters to determining a way in which nature herself produced stars of this sort, the matter nevertheless relates to God, the author of nature; and even when the emergence of the star [of 1604] was attributed to ordinary nature, there still remained a place for God to arrange the circumstances of time and place.[67]

Along with the unforeseeable nature of divine providence, Kepler claimed that celestial novelties such as the new star of 1604 held an enormous significance for the future of humanity. "Such a congruence of place and time," Kepler wrote, could not "be attributed to anyone but God alone."[68] That God "employed an ordinary way of effecting this new star," Kepler claimed, did not refute the fact that it expressed "an eminently good end, at which providence aims."[69] Although the meaning of this message may not have been immediately clear, the principal recipient was all but indisputable. Seen from the earth, the new star was positioned closely to the conjunction of the superior planets. From any other perspective, it possessed no apparent connection to them at all. "For what good," Kepler asked, "did God adapt the star to those things that appeared not in the highest and most vast ether but here on this humble and lowly earth?"[70] As Kepler explained, nature had conveyed a divine message through this spectacle to the favorite creature of the cosmos. Kepler believed he had made this clear in his refutation of the "first opinion" of astrologers who claimed that the conjunction of the three superior planets had in some way created the new star:

> For which creature do you suppose that this spectacle was displayed? For some celestial being? In the refutation of the first opinion, however, it was said that nowhere in the world did the star appear associated with the planets except from the earth. And so could this spectacle have really been displayed for any other terrestrial creature than man?[71]

---

67 Ibid., 284.14–18: "Nam etsi per superiora aliquot capita hoc egi, ut modum invenirem, quo Natura ipsa ederet huiusmodi sidera: res tamen ad Deum redit, naturae authorem; et ipso sideris ortu naturae ordinariae transcripto, relinquebatur tamen locus Deo, in concinnandis loci et temporis circumstantiis."

68 Ibid., 288.6–7.

69 Ibid., 287.39–288.11.

70 Ibid., 289.30–32.

71 Ibid., 287.40–288.5: "Cui enim creaturae propositum existimas hoc spectaculum? Num alicui coelesti? At dictum est in primae sententiae refutatione, nuspiam Mundi, nisi

Kepler regarded the meaning of these celestial messages as something beyond the realm of prediction and understood only retrospectively. That their significance was not immediately clear allowed for an examination of conscience, he claimed, and an opportunity for introspection in anticipation of extraordinary events. As Thomas Mingonius wrote to Kepler in a letter of 1619, "Of course, we should all pray to God that such signs of His anger may be empty or at least more lenient."[72] Whether or not Kepler thought human appeals could alter the course of providence, he suggested that they encouraged practices that could only improve the spiritual condition of humanity. With the growth of appreciation came a greater regard for God's creation, which was thought to have been established for the physical and spiritual well being of humanity in the first place. The unpredictable nature of celestial novelties did not imply that their material causes and consequences on earth could not be grasped in some way by the human mind and spirit. As the primary cause of the cosmos, God was held accountable for the order of nature and the acquisition of human knowledge. "God reveals deep and secret things," Kepler wrote, recalling the words of Daniel, and "He knows what is in darkness; the light dwells with Him."[73]

Rather than interpret the meaning of celestial spectacles straight away, Kepler allowed for the passage of time to permit a better perspective on the divine plan. According to theological authorities such as Heerbrand, comets were the substantiation of providence and had "something to tell the observer about the will of God."[74] Without understanding the importance of these events immediately, human observers were considered more capable of comprehending them with time, in some cases retrospectively as the warnings of particular circumstances. As one example, Kepler suggested the story of Sebastian I (1554–1578), King of Portugal, whose disastrous defeat at the hands of the Turks Kepler attributed to the stimulus felt by the comet of 1577. Despite not receiving support promised by his uncle, King Philip II of Spain (1527–1598), Sebastian "crossed into enemy territory" in Morocco with an army consisting mainly of mercenaries.[75] Short of experienced soldiers and against

---

ex sola terra, visum esse sidus Planetis associatum. Num igitur alii alicui terrestri creaturae propositum est hoc spectaculum, praeter hominem?"

[72] *JKGW*, 17, no. 821, 8–10.

[73] *JKGW*, 1, 292.8–9; cf. *Daniel*, 2.20–22.

[74] Methuen, 1999, p. 107.

[75] *JKGW*, 8, 230.33; *JKGW*, 4, 62.42.

the wishes of his advisers, Sebastian "perished by his own fault along with his army."[76] What Kepler saw as significant about this story was neither the natural temperament of Sebastian, described as stubborn, "blinded by desire," and "doubtful of the opinion of the more discerning,"[77] nor the emotional stimulus the comet gave him. Rather, Kepler stressed the significance of the comet as a celestial messenger that was later revealed as a warning against the disastrous consequences of allowing one's emotions to determine a decision. "Free reason had been given by God to [Sebastian] to tame the passions," Kepler wrote, and "he could have done so if he had wished."[78] Warning the king of "imminent disaster from blind emotion," the comet had been wrongly read by Sebastian and led to his ruin.[79] In allowing his emotions rather than reason to determine his aggressive military maneuver, Sebastian had brought about the very thing the comet could have come to discourage.

Another role that Kepler thought comets could play was their part in announcing the births of extraordinary individuals. Kepler entertained the idea that the widespread visibility of comets and their capacity to move human witnesses could in some way reflect the nature of those persons of renown born around the time of their appearance. "What in the heavens is a new and wandering comet," Kepler wrote, "may on earth be some new man" who held sway over others.[80] In the same way that a comet struck fear in the hearts of many, those individuals born under their appearance could win renown "either by the magnitude of authority or force, or by stirring up some special fame through their teaching."[81] Among those born around the appearance of a comet was Alexander the Great (356–323 B.C.E.), who, "born in a year in which Pliny recalls there was a comet, inflicted the greatest ruin on the human race."[82] Kepler also referred to Mithridates (132–63 B.C.E.), whose birth and ascension to the throne of Pontus were accompanied by the appearances of "terrifying comets."[83] In either case, a comet had warned of the egregious nature of an important person, providing a forerunner of infamous acts whose realization only required the passage of time.

---

76 *JKGW*, 8, 230.32–34; *JKGW*, 4, 62.42–43.
77 *JKGW*, 8, 230.32–38; *JKGW*, 4, 62.40–41.
78 *JKGW*, 8, 230.38–40; *JKGW*, 4, 63.6–8.
79 *JKGW*, 8, 230.40–231.2; *JKGW*, 4, 63.7–10.
80 *JKGW*, 1, 341.5–6.
81 Ibid., 341.6–9.
82 Ibid., 341.9–11.
83 Ibid., 341.11–12.

In his view of comets as retrospectively revealing, Kepler argued that the astrological prediction of events was done "at the risk of one's own faith."[84] How the circumstances of the comet of 1607 served as a source of influence on earth, he said, was "a slippery affair."[85] Kepler promised little when it came to his own prognostication, even though he knew the practice of prophecy well. Kepler often claimed that divine providence did not allow for the forecasting of future events. Heresy would take on the form of a prediction that presumed the theoretical resources of astrology could reveal the meaning of celestial messages prematurely. Kepler denied that the path of the comet of 1607 in the direction of the site of the new star of 1604 suggested it would relate to the star and set a course for what was thought to be signified by it, namely "the reformation of the world."[86] Kepler also refused to interpret anything in terms of individual events, since he supposed that the heavens did not articulate any single circumstance.[87] In any case, Kepler took up his "office as astronomer" and kept his distance from the task of prophecy.[88] He pointed in particular to "the impiety, folly, and foolishness" of horary astrology, declining to deal with it in any way.[89] Kepler saw such activities as "establishing a new empire, transforming kingdoms and principalities, and managing mutations of religion" as the outcome of human enterprise, knowable only to "inscrutable divine providence."[90] To conjecture on the influence of comets in the political and theological spheres was simply beyond the ability of human beings. Kepler claimed the only element of truth in this form of conjecture spoke to our fundamental nature and not the constitution of the heavens. As an example, Kepler referred to a poetic account of a comet where he had found a simple truth expressed about the heretical nature of human beings:

> I once saw a poetic account of a comet in which the comet with its long tail was very elegantly compared with a great heretic who would seduce the people with fanatical speeches. I certainly do not reject such an explanation,

[84] *JKGW*, 8, 244.25.
[85] Ibid., 244.24–25.
[86] Ibid., 244.37–40; cf. *JKGW*, 4, 72.22–25.
[87] Rabin, 2010, p. 62: "Kepler believed in astrology, but he was not an astrological determinist."
[88] *JKGW*, 8, 244.25.
[89] *JKGW*, 8, 248.6–14; cf. *JKGW*, 4, 74.41–75.4.
[90] *JKGW*, 1, 326.26–29.

> unless God averts this omen, as I pray; for heresy is everywhere, and there is no one who can deny that he has not attempted it.[91]

Recalling the words of Christ, Kepler wrote that the Son of God had spoken of "signs among the stars and distress of the people on earth."[92] Considering the amount of confusion and excitement a comet could cause, Kepler may have seen them as an opportunity to act prudently and never lose sight of our moral substance before the End of Days. The response to these celestial announcements lay in the responsibility of the individual who decided either to endure the impending disaster, attempt to avoid it, or suffer it "with a more virtuous spirit."[93] The latter option fulfilled the proper part of providence that encouraged the spiritual progress of humanity. In this way, Kepler wrote, "the pretences of religion, conscience, and piety may be burned in the fire of temptation, and the desires and wrongdoings lurking beneath these pretences may be laid bare and the causes of sin ascertained."[94] In the light of a new luminary, "the persistence of man towards an evil purpose" could finally be turned in the right direction and the abuses of political rule drowned "in a mighty sea of misery and tears."[95]

Among other scenarios, Kepler suggested that the brevity of the comet of 1607 may have also been testimony to the mortality "of each and every one of us."[96] Observed by few and cut short by the particular perspective of the earth, the comet could have served as a stark reminder of "the shared condition of the human race."[97] This condition gave no privilege "to the strong over the weak," Kepler wrote, nor "the young over the old," and it certainly awarded no advantage to "the prophet over the counselors."[98] That the comet offered further evidence of the mutability of the cosmos was sufficiently clear. Kepler believed that the linear motion of comets evinced their transient nature. He claimed they confirmed that "the entire

[91] *JKGW*, 8, 245.16–20: "Vidi olim descriptionem Cometae poëticam, in qua Cometa cum longa sua cauda perquam eleganter comparabatur Haeretico novo, qui fanaticis concionibus populum seduceret. Sanè nec hanc rejicio explicationem, nisi quòd, ut omen Deus avertat, precor; haeresium enim abundè est, nemo contra, qui haeresin intentatam fateatur."

[92] *JKGW*, 8, 239.36; cf. *Luke*, 21.25.

[93] *JKGW*, 8, 261.9–11.

[94] Ibid., 261.18–21.

[95] Ibid., 261.13–18.

[96] Ibid., 238.31.

[97] Ibid., 238.32–33.

[98] Ibid., 238.34–36; cf. *JKGW*, 4, 67.2–5.

universe, all the way up to the heavens themselves, is corruptible and (in parts, at least) changeable over the course of time."[99] This reality spoke to the mortality of man. "The condition of his mortality," Kepler declared, "must be considered by man with equanimity." Comets were just the right occasion to contemplate one's own expiration more acutely.[100] While the comets of 1618 further confirmed this view, they had also inspired others to put forward a more catastrophic forecast. Their retrograde motion had been seen by some as suggesting political unrest, "adverse in every way to the well being and wishes of the populace."[101] Their motion in the direction of "the sides of the world," meanwhile, had persuaded others to predict "the incursion and plundering of barbarians and the oppression of the Christian religion," among other things.[102]

Again, there was no mistaking the recipient of the providential message of comets. Kepler claimed that they relied on natural causes that would rarely result in their appearance to human observers on earth. Through his study of celestial novelties, confirming in the case of the new star of 1604 that an "assembly with the conjoined planets" could only have been perceived from our perspective,[103] Kepler underscored the anthropocentrism of his cosmology. Although he admitted that countless things were constantly occurring according to their own natural causes, Kepler perceived these causes as the product of providence expressed in a way that human observers were capable of contemplating, if not fully comprehending.

Kepler denied that the comets of 1618 had anything to do with the two years of poor health preceding the death of Holy Roman Emperor Matthias (1557–1619). "There was no need for the comet to indicate this in advance," Kepler said specifically of the comet that had appeared in the sign of Scorpio, Matthias's ascendant, "since the future could be foreseen by the laws of nature."[104] Kepler remained hopeful, however, that the second two comets that had appeared in November 1618 would sway the populace to a speedy resolution of the Thirty Years War. Rather than refer to the conflict directly, Kepler recalled an example from history that saw civil war in ancient Judea quelled by the arrival of Pompey and the Roman army in 66–64 B.C.E. Kepler sensed in this example the presence

[99] *JKGW*, 8, 239.10–12. Cf. Ruffner, 1971.
[100] *JKGW*, 8, 239.12–16.
[101] Ibid., 256.29–31.
[102] Ibid., 256.30–35.
[103] *JKGW*, 1, 289.32–35.
[104] *JKGW*, 8, 259.28–30.

of divine providence, whereby God, "upon the arrival of Pompey with the Roman army, dissolved that war in such a way that neither side would win, but both would serve the Romans from that time."[105] Kepler clearly hoped that providence and the progress of the Holy Roman Empire would be similarly served by the swift conclusion of the Thirty Years War.

Clarifying Curvature and the Rectilinear Course of Comets

Kepler argued that the location and motion of a comet could only be found after first determining the observational effects of the motion of the earth. Nothing could be done to track the comet of 1607 when it disappeared "in the hemisphere of the Sun" during the final days of September, Kepler wrote, "except by supposing the motion of the earth."[106] This motion accounted for the retrogradation of the comet as well as the early days of displacement, which were explained by the fact that "the diurnal motion of the comet at the beginning of its trajectory matched in longitude the daily motion of the earth."[107] Finally, the tail of the comet had disappeared so quickly not because it had actually done so, but because the comet had moved towards the Sun on a line that "faced obliquely away from the earth,"[108] impeding observability as soon as it reached a certain point.

In his account of the second comet that appeared in 1618, Kepler recalled the report of imperial physician Johannes Remus Quietanus, who had described "the figure of an ostrich feather," curved like the sort "that the ancients called horn-shaped [*ceratias*]."[109] Remus, who had also given Kepler an account of "a burning beam [*trabs*]" observed in Rome, was just one of several sources Kepler employed in his broad empirical repertoire.[110] Kepler also drew from his own experience, as he did when describing his final observation of the second comet. Among the black clouds and strong winds that filled the sky on an early morning near the end of November, Kepler glimpsed the comet, "greatly dissipated," one last time:

105 Ibid., 262.8–10.
106 Ibid., 178.20–23.
107 Ibid., 178.23–24.
108 Ibid., 178.24–25.
109 Ibid., 180.19–20.
110 Ibid., 180.20–21.

> On Thursday, 29 November, at 5:00 in the morning, when the sky was found among black clouds and strong winds and the plains were painted with snow, only the trail of the comet still appeared, greatly dissipated and outdone in whiteness by the clouds illuminated by the moon. At that time, it stretched beneath the stars of the southernmost coil of Hydra... This was the final observation of the second comet.[111]

On the same day the second comet of 1618 disappeared, Kepler observed a third comet. As he recalled in his own account, the latter comet appeared very brightly "below the left scale of Libra," with a tail scattering in straight lines:

> On the same day, 29 November, at 6:30, when the clouds dispersed again I went up to the roof to see if perhaps something could be observed more accurately in the tail of that [second] comet, but it had already disappeared with the break of day... Another, very bright comet appeared from the clouds, however... The color of the tail twinkled yellow and red in such a way that it was long at one time and short at another, scattering from the head like stiff and straight brooms.[112]

More enduring than the comet of 1607 and the second comet of 1618, the third comet remained visible for well over a month. "The head and tail," Kepler recalled, shone brilliantly "for all of December and a good part of January."[113] The date of the disappearance of the comet was disputed. Ambrosius Rhodius (1577–1635), a professor of mathematics at Wittenberg, claimed that the comet disappeared on 2/12 January "near the penultimate star in the tail of Draco."[114] Wilhelm Schickard (1592–1635), on the other hand, had given Kepler an illustration that extended the life of the comet to 8/18 January.

In May 1619, Kepler completed what he called "a three-part treatise on comets,"[115] with the first and astronomical part introducing "new and

---

[111] Ibid., 185.27–35: "Die 29. Nouemb. mane hor. 5. cùm coelum ad momentum detegeretur inter atras nubes et ventos vehementes, cùm campi essent picti nivula, apparuit tamen tractus iste... Cometae, sed valdè dilutus, nec aequans albedinem nubium à Luna illuminatarum: Iam porrigebatur infra stellas spirae Hydrae australissimae... Haec ultima Secundi Cometae fuit obseruatio."

[112] Ibid., 185.37–186.6: "Eodem die 29 Nouembris hora 6 $^1/_2$, cùm rursum discuterentur nubes, ascendi tectum, si fortè accuratius aliquid in illa Cometica cauda veniret notandum: Verùm illa iam disparverat adulto divuculo... at... Comet alius clarissimus per nubes apparuit... Color caudae inter flavum et rubrum micabat, ut interdum longa, interdum breuis esset: spargebatur à capite, ut scopae directae et rigentes."

[113] Ibid., 177.9–11.

[114] Ibid., 196.18–20.

[115] *JKGW*, 17, no. 862, 22–24.

remarkable theorems" explaining the paths of motion of the comets of 1607 and 1618.[116] In the second and "physical" part, Kepler proposed "a new and extraordinary physiology [*physiologia*] of comets," and in the third part he gave an analysis of the astrological influence of the three comets.[117] Published in 1619, *On Comets* was proudly adorned with a frontispiece featuring a passage from the *Natural Questions*, in which Seneca had anticipated someone who would one day demonstrate "into what parts comets wander, why they advance so far away from each other, how many there are, and of what sort."[118] Kepler seems to suggest he is the suitable successor who can finally answer the questions that Seneca had once asked. Again, Kepler argued that the motion of the earth played an essential part in making sense of cometary motion. He condemned the curvature of their course as "a mere deception of vision, like the stations and retrogradations of the planets."[119] Only by taking into account our terrestrial perspective did Kepler demonstrate, for example, the elevation of a comet and "how it followed a fully uniform trajectory in a straight line."[120]

Kepler foresaw opposition to *On Comets*. Not only did his account build on a heliocentric astronomy, but it introduced a controversial form of motion into the heavens. Kepler claimed that comets, in contrast with the planets and their periodic paths, traversed a straight-line trajectory across the celestial ether until eventually evaporating. Seen in this way, comets and planets expressed a particular pattern of motion depending on their durability. In the case of comets, rectilinear motion reflected transience and terminability, while the circular motion of planets signified perpetuity. Kepler expressed this as a relation between the condition of a body and a corresponding form of motion:

> And since a difference between timeless and transitory bodies is also observed by a difference in their motions, circular [motions] and certainly revolutions shall be of perpetual bodies . . . while rectilinear [motions] shall be of evanescent ones. For either one possesses the cause of its own condition in its own particular form: [the condition] of eternity in the circle [and the condition] of mortality in the straight line, which can certainly not be infinite.[121]

---

116 *JKGW*, 8, 131.3–6.
117 Ibid., 131.7–10.
118 Ibid., 131.15–18; cf. Seneca, 7.25.
119 *JKGW*, 8, 219.33–34.
120 *JKGW*, 17, no. 831, 8–9. See Solís, 2001, p. 50.
121 *JKGW*, 8, 217.3–8: "Et cùm differentia perennium temporariorumque corporum sequatur in eorundem etiam motus, circulares utique revolutiones erunt perennium

Since straight-line courses could only continue so far in Kepler's finite cosmos, comets were necessarily corruptible. In turn, Kepler viewed their rectilinear motion as evidence of the material essence of the heavens, whose constitution he claimed was "fluid and everywhere penetrable."[122]

Kepler feared that *On Comets* would reach fewer readers on account of these controversial claims. In a letter to Vincenzo Bianchi of 1620, Kepler wrote that his book would "scarcely be sold in Italy," where censure would quickly follow on the heels of his attempt to "save the appearances of a comet by its uniform rectilinear trajectory in the ether and by the earth's motion around the sun."[123] Kepler had good reason for his doubt. While Bianchi had encouraged Kepler to see to it that "Italy draw pleasure and utility" from his works in the same way that other countries enjoyed them, Bianchi was well aware that hostility towards Copernicus was still fresh and the source of "fervent censure" there.[124] When Galileo sought a copy of the first part of Kepler's *Epitome of Copernican Astronomy* (1618–1621), he had requested it from the royal court in Linz rather than Florence, where it was prohibited. Bianchi suggested another way of selling *On Comets* by making it available secretly in the same way as other German authors who were formally forbidden in Italy. "Whenever the books of famous German authors are forbidden," Bianchi wrote, "they are still sold secretly and read more diligently."[125] Bianchi praised the erudition of Kepler and claimed that the rarity of his works in Italy only added to their worth:

> Strange to say, but your works are found neither in Venice nor Rome nor any other prosperous city in Italy unless by great effort or good fortune ... If a thousand books of yours were sent here, they would all be sold at a high price.[126]

Bianchi compared Kepler's reputation in Italy to the image of a Phoenix, rare in appearance yet all the more admired as "the greatest mathematician

---

corporum ... rectilinei verò vanescentium: habent enim utrique conditionis suae causam in forma quisque sua: aeternitatis in circulo, mortalitatis in linea recta, quae infinita utique nequit esse." Cf. Ruffner, 1971, p. 181. It is important to note Kepler's Christian view of perpetuity, which was qualified by the Second Coming. See *JKGW*, 1, 351.14–15.

122 *JKGW*, 8, 220.6–7.

123 *JKGW*, 17, no. 862, 22–24.

124 *JKGW*, 17, no. 825, 109–111.

125 Ibid., 111–113.

126 Ibid., 103–107: "Mirum dictu, nec Venetiis, nec Romae, nec florentibus aliis in urbibus Italiae, Opera tua nisi maximo labore, vel casu reperiri ... Huc si tui ... mille transmitterentur libri, magno omnes venderentur pretio."

in all of Germany."[127] For his own part, Kepler continued to look to Italy with caution.

Kepler called on the Pythagoreans in his case against Aristotle, who had argued that comets were set on fire in the upper atmosphere and progressed in perfect circular motion around the earth, "the center of the universe."[128] Although he did not completely agree with the Pythagorean view of comets, Kepler accepted their explanation of how they disappeared when they approached the sun. Considering "the constant experience of comets concealed by the rays of the sun," Kepler recalled, the Pythagoreans had denied that "comets are extinguished, but by a certain principle they are received by the Sun and then proceed onward again."[129]

Remus, among others, assigned a more active role to the sun. In the summer of 1619, he argued in a letter to Kepler that the sun played a pivotal role in the generation and appearance of comets. A comet was either an object "collected from sunspots and compacted together," Remus wrote, or "frozen air above the sphere (that is, the path) of Saturn," where the solar rays were rather weak, as witnessed by "the pallor of Saturn and the poor illumination of the sun" there.[130] Just as he identified the sun as a factor in the formation of comets, Remus supposed that the sun determined their discernibility. If comets were made up of air, he reasoned, and the air was thinner "near Mercury and Venus than Saturn," the aerial substance of comets above the planets would provide "a wall" of dense matter and a good place to illuminate.[131] Remus did not believe that this reception of light rays led to their refraction when "crossing through the cometary head," and he suggested they rarefied comets "by the force of heat."[132] "You see that I do not appoint an imaginary body or a mere reflection," Remus wrote, as a comet would melt into a spherical or elliptical shape according to what he saw as "the nature of celestial bodies."[133] As for the actual course of a comet, Remus claimed that he could prove their curved path in concert with the motion of the earth. "Would that your book appear," he wrote to Kepler on the eve of *On Comets*,

---

127 *JKGW*, 17, no. 724, 42–44.
128 *JKGW*, 8, 276.5–10.
129 Ibid., 219.10–15.
130 *JKGW*, 17, no. 848, 38–42.
131 Ibid., 42–44.
132 Ibid., 46–52.
133 Ibid., 52–56.

"for I observed a curved path that I found could be firmly defended according to the motion of the earth."[134]

Although Kepler would employ the observations of Remus, he did not accept the conclusions Remus had reached concerning the curvature of cometary motion. Kepler explained away any deviation from a steady and straight-line path by accounting for the motion of the earth. Either the comet drew so far away from the earth that its displacement was scarcely discernible, or the motion of the earth matched that of the comet in the same direction, or the comet even appeared to reverse course as a consequence of the earth's greater swiftness. When asking the question "why comets begin to curve their courses when they slow down,"[135] Kepler offered an explanation along the same lines. While he had "once thought it was real," Kepler now diagnosed this form of motion as an optical illusion for which he gave an account before establishing the true motion:

> I once thought it was real, and so I philosophized about the physical causes. But it is a mere deception of vision, like both the stations and retrogradations of the planets. For a comet receding in a straight line far away from the earth greatly reduces its latitude and does not make any apparent longitudinal progress ... The motion of the earth becomes strongly sensible when reckoning the swiftness and location of comets.[136]

As "a mere deception of vision," the curvilinear course of comets would serve to support Kepler's heliocentric view of the cosmos. Convinced that the heavens were filled with a countless number of comets and that only a select few had ever been seen, Kepler claimed that "there are as many comets in the heavens as there are arguments that the earth is moved in an annual motion around the sun."[137] By attributing their paths to the motive power of the sun, Kepler also put forward the possibility that comets were moved by the same source of motion as the planets. In particular, Kepler suggested that comets sped up and slowed down in the same way that the planets did depending on their direction and distance from the sun. Seen as the central magnet and motor of the universe, the sun could

---

[134] Ibid., 63–65.

[135] *JKGW*, 8, 219.30.

[136] Ibid., 219.32–37: "Olim existimavi esse reale quippiam; itaque de causis Physicis philosophabar. Sed est mera visus deceptio, ut et stationes et regressiones Planetarum. Cometa enim in recta linea multum à Terra recedens, multum minuit latitudinem, et non facit evidentem progressum longitudinis ... motus Terrae celeritatis et situs ratione in Cometis ... valde fit sensibilis ..." Cf. Ruffner, 1971, p. 180.

[137] *JKGW*, 8, 220.17–19.

apparently affect the motion of comets in the same way that it acted on the planets:

> The natural philosopher shall find that some comets, like the one in 1607, are less inclined to motion early on, and so they perhaps rarefy at the same rate as they move; while other comets languish at the end, as in the case of the comet of 1618. Let the examiner consider whether the cause might be that the former [comet] was carried towards the sun and the latter [comet] away from it. For it so happens in this way with the circular motion of the planets around the sun, as they increase or decrease their motion according to their rectilinear proximity to or removal from the sun.[138]

As this passage suggests, Kepler was convinced that comets evinced the centrality and motive influence of the sun. "Farewell, Ptolemy," he wrote, "I return to Aristarchus with Copernicus as my guide."[139]

Celestial Sympathy and Earthly Knowledge of Comets

In the second part of *On Comets*, introduced as an edition of the report "written twelve years ago in Prague and now newly reconsidered,"[140] Kepler examined the natural causes of comets more closely. Comparing comets to airborne projectiles, Kepler viewed their material essence as in many ways the same as the substance of air surrounding the earth. Comets were not fissures, flames, or fireworks, but they did follow paths that resembled projectiles, and they appeared to rarefy under the influence of the sun in the same way as the clouds. In his comprehensive cosmology, Kepler conceived of comets and everyday events in the sublunar sphere according to the same material experience. Comparison with terrestrial objects allowed Kepler to express their behavioral similarities while arguing for their many forms of physical continuity.

In a brief but substantial study entitled *The Six-Cornered Snowflake* (1611), Kepler had spoken of a natural faculty that served to realize archetypal principles in all of material existence. These principles were geometrical

---

[138] Ibid., 220.12–16: "... discet Physicus, Cometarum aliquos, ut illum anni 1607. inter initia minus aptos esse ad motum, itaque fortassis attenuari eadem proportione qua et moventur; aliquos in fine potiùs languescere, ut in Cometa anni 1618. accidit. Cogitet speculator an causa sit, quod ille versus Solem est delatus, hic à Sole. Nam sic fit in circulari motu Planetarum circa Solem, ut illi motus suos intendant vel remittant pro ratione propinquitatis vel remotionis à Sole rectilineae."

[139] Ibid., 220.19–20.

[140] Ibid., 221.6–7.

figures that found their original form "in the mind of the Creator," Kepler claimed, and were "coeternal to God."[141] The presence of these principles could be proven by many examples, Kepler wrote, including the pentagonal form of flowers and the hexagonal shape of honeycombs, "the most spacious shape for storing honey."[142] On earth, the natural faculty was "one and the same," and yet it broke up, as it were, into different bodies and acted on them "according to the internal condition of their matter" along with "the external circumstances."[143] If this same universal faculty could account for the hexagonal shape of snowflakes, it could by analogy also account for the archetypal essence of every natural phenomenon, including comets.

Helisaeus Roeslin, who claimed that the comet of 1577 had traced a path in the spherical orb of Mercury, attributed cometary motion to a mathematical principle conceived "according to considerations concerning the reciprocal harmony and symmetry" of comets.[144] Roeslin was convinced of the immutability of the heavens and set comets apart as "miraculously created by God."[145] According to this view, the short life of comets signaled "nothing other than the imminence of the end of the world."[146] Kepler, by contrast, thought the path of a comet was measured and ordered by a geometrical principle and a natural cause that he associated with his polyhedral hypothesis and later harmonic theory. As with the construction of a honeycomb, where the hexagon was employed by a bee according to "an archetype stamped on it by the Creator,"[147] comets followed a course according to architectonic principles that were realized by a natural faculty and recalled the essence of their divine author. Although Kepler spoke critically of the spiritual principles that Thales, Pythagoras, and other ancient authorities had identified in the heavens, he also suggested the possibility that a comet was led by "an invisible spirit" to approach the earth.[148] It was precisely this sort of speculation that had stirred the opposition of the theological faculty at Leipzig twelve years earlier.

---

141 *JKGW*, 4, 276.35–36; cf. Kepler, 2005, p. 48.
142 *JKGW*, 4, 269.24–26; cf. Kepler, 2005, p. 40.
143 *JKGW*, 4, 276.17–20; cf. Kepler, 2005, p. 48.
144 Granada, 1997, p. 410.
145 Ibid., p. 411.
146 Ibid., p. 411.
147 *JKGW*, 4, 269.12–15.
148 *JKGW*, 8, 233.6.

Kepler suggested that comets, like new stars, arose from dense areas of the ether. He compared them to "a sort of abscess or excrement,"[149] expressing their role in the cycle of condensation and cleansing that the ether constantly underwent. When the ether thickened in a particular part, Kepler explained, it blocked the light of the sun and the stars in such a way that created a "need for purging and purification." Incited by the sun, a comet was a sign and symptom of this process. Kepler claimed that it was the product of a natural faculty that lay in "the very substance" of the ether and acted in the same way as an animate or vital faculty:

> [A comet] often occurs whenever the ethereal aura condenses in certain regions, and it is for this reason that the light of the sun and the stars does not freely reach the earth. Thus, history tells us that whenever the sun is covered by a red cloak the color of blood or rust, either for several days (as in the four-day period of 24–28 April 1547), or for a full year (as in the year of Caesar's death), there is then need for purging and purification. The faculty that lies in the very substance of the ethereal aura fulfills this in the same way as an animate or vital faculty.[150]

Despite their origin in the heavens, Kepler claimed that comets moved in straight lines "like meteors or fireworks" and not in a circular path like "the [motion] of the perpetual planets."[151] He explained that their eventual endpoint, "whether they are extinguished, break apart, or burn out and become ashes,"[152] was not entirely clear because they often disappeared beneath the rays of the sun before vanishing completely. However they came to an end, Kepler believed that comets confirmed there was "no less generation and corruption in the heavens" than in the air on earth.[153] As for "that ancient opinion of Aristotle," who assigned the material origin of comets to sublunar vapors, Kepler criticized his own contemporaries for reiterating this view "with so much enthusiasm and defending it with so much ardor."[154] As if arguing their case purely "by the prescription of

---

[149] Ibid., 225.18–19.

[150] *JKGW*, 8, 225.9–19: "Solet quandoque aura aetherea crassescere certis regionibus, eoque fit, ut stellarum Solisque lumina non liberè perveniant ad Terram. Sic testantur Historiae, Solem quandoque per dies aliquot, ut anno 1547. quatriduo à 24. in 28. Aprilis, quandoque per integrum annum, ut anno caedis Caesaris, palliolo, ferrugineo, sanguineove colore hebetari: Tunc igitur defaecatione et purgatione opus est, quam praestat facultas illa, quae inest in ipsa substantia aurae aethereae, similis animali aut vitali facultati." Cf. *JKGW*, 4, 59.11–18.

[151] Ibid., 59.35–37; cf. *JKGW*, 8, 226.1–3.

[152] Ibid., 226.11–12; cf. *JKGW*, 4, 60.3–4.

[153] *JKGW*, 8, 225.28–32; cf. *JKGW*, 4, 59.21–26.

[154] *JKGW*, 8, 225.35–37.

antiquity," Kepler wrote, they ignored experience and reduced their task to merely a mental exercise.[155]

Faced with accounting for the tails of comets, Kepler claimed that the sun played a pivotal role in their development, direction, and duration.[156] On first encountering the spherical surface of a comet, the rays of the sun were thought to refract and continue through it, straining and attenuating the comet in such a way that would produce a tail. Kepler compared this process to a thread of silk that would eventually exhaust the body of the silkworm:

> When the matter [of a comet] is condensed into a sphere according to the nature of all things that are drawn together, and the rectilinear rays of the sun strike and penetrate this pellucid globe, I believe that something proceeds from the inner matter of the comet and extends along the same path by which the rays of the sun break through; and in this way, the body of a comet is bathed, strained, attenuated, and finally annihilated. Like a thread cast by a silkworm, comets are thus consumed and eventually die by discharging a tail.[157]

The emanation of rectilinear light rays from the central body of the sun explained why the tails of comets always extended away from the sun. Material from the body of the comet, Kepler argued, was "continually expelled by the rays of the sun" in the other direction.[158] This explained why the tail of a comet "always extends in nearly the opposite direction of the sun," Kepler wrote, and why the rays of the sun became visible in the ether when they encountered the dense body of a comet.[159] Comets were illuminated by the sun and made to discharge their own substance, gradually dissipating into the same ethereal vapor from which they originally arose. Kepler compared the attenuation of comets to the evaporation of clouds, whose substance was measured as "four times greater than the matter of comets."[160] Although clouds were effected by the sun in different ways, with one part condensed and cast "dripping down onto the

---

155 Ibid., 225.37–39.

156 For an overview of Kepler's cometary theory over the course of his career, see Barker, 1993, pp. 19–22.

157 *JKGW*, 8, 226.13–18: "Coacta materia in orbem ex natura omnium rerum quae uniuntur, et Solis rectis radiis pellucidum hunc globum ferientibus atque penetrantibus, existimo sequi aliquid de intima Cometae materia, exireque viam eandem, qua perrumpunt Solis radii; atque hoc pacto corpus Cometae perlui, colari, atteri, et denique annihilari, et sicut bombyces filo fundendo, sic Cometas cauda expiranda consumi et denique mori."

158 Ibid., 226.28–29.

159 Ibid., 226.29–34.

160 Ibid., 226.22–25.

earth" and the other "dissolved and raised up on high,"[161] Kepler offered their diffusion as an example of how comets might be similarly rarified and returned to the ether.

Kepler next explained why the tails of comets were often seen as curved. He claimed that he had already shown that the rays of luminous objects such as the sun were not arched, and so there had to be another reason for their curvature. "Although the tails of some comets are curved," Kepler wrote, "it is impossible that the rays of the sun are curved in the open ether. For as we learned in the *Optical Part of Astronomy* [1604], the rays of luminous objects are not curved but straight lines."[162] Kepler attributed any "curvature of the tail away from a straight line in opposition to the sun" to an optical effect rather than the solar rays.[163] As an example, he referred to the second comet of November 1618, the end of whose tail had displayed an angle the head no longer held. Trailing behind, the tail end of this comet had kept a course that the head, the first part to be influenced by the sun, no longer preserved. "Imagine that the swiftness of the head is so great," Kepler wrote, "that the tail is left behind, especially near the end."[164] Kepler called this phenomenon "a fiction," implying it was the result of our vision and not a reflection of reality.[165]

As for why a comet's tail deflected away from direct opposition to the sun, Kepler suggested the possibility that "the surface of the cometary head was uneven."[166] In the case of the comet of 1577, this had led to the manufacture of multiple tails streaming away from the surface, each one "by a certain and separate measure."[167] The comet of 1607, on the other hand, had suddenly grown long "in the blink of an eye," spreading out to the sides like "a flame," while the head of the comet had continued forward.[168] The partition of comets was also a possibility that Kepler considered for the first comet of November 1618, which disappeared on the same day that the second comet was first observed. Whether or not the two comets were related awaited the assembly of observations "made

161 Ibid., 226.25–27.

162 Ibid., 226.38–40. On the principle of punctiform analysis in Kepler's optics, see Lindberg, 1976, p. 193.

163 *JKGW*, 8, 227.3–5.

164 Ibid., 227.10–14.

165 Ibid., 227.4.

166 Ibid., 227.29–30.

167 Ibid., 227.30–31.

168 Ibid., 227.40–228.4.

from around the globe," Kepler wrote, "especially if those sailing across the Indian Ocean had paid careful attention."[169]

Despite his attempt to employ a wide range of earthly analogies, Kepler did not accept the accounts of others who simply equated comets to "flames, flickering to the sides" and "stirred by winds."[170] Kepler claimed that comets were not fiery objects because their tails did not glow red and their bodies could not have been set on fire by the sun. Their tails extended "constantly in the same direction," he wrote, "away from the sun."[171] And although their illumination was entirely the sun's making, their strands of hair, swerving ever so slightly, were not flames sustained by the material substance of the head but a rarefied form of it projecting in the distance. That some comets became visible in the first place, however, was something so remarkably rare that Kepler remained reluctant to attribute their appearance entirely to natural causes. Nevertheless, Kepler spoke of "a natural connection" between a comet and "an internal faculty of the earth" when he turned to the task of predicting the weather.[172]

Kepler rejected many of the natural effects on the weather others claimed for comets. Several of the philosophers followed by the astrologers, "especially Aristotle," had relied on two false foundations for their theory, Kepler claimed, (1) that comets were fiery objects and (2) that they existed below the moon.[173] Along with these two claims, Kepler wrote, came "other doubtful and false things."[174] Citing evidence that suggested certain events were more prone to take place, Kepler identified the existence of a natural faculty in the earth that brought on conditions that were not normally witnessed without a comet. "The true way in which comets alter the state of the air," Kepler wrote, required the active presence of "a natural faculty in the earth that may effect such [meteorological] events that we attribute in vain in the absence of a comet."[175] The drought and dryness often feared after the appearance of a comet were now thought to come from a faculty that produced these conditions whenever it was "stimulated by a new phenomenon" in the heavens.[176] The earth could react in other ways, however, as Kepler recalled the massive amount of

---

[169] Ibid., 179.38–180.1.
[170] Ibid., 227.32–35; 228.33–35.
[171] Ibid., 228.34–35.
[172] Ibid., 238.6–15; cf. *JKGW*, 4, 66.25–27.
[173] *JKGW*, 8, 229.20–39.
[174] Ibid., 229.40.
[175] Ibid., 230.1–5.
[176] Ibid., 238.14–17.

snow that fell after the comets of 1618. "With dryness prevailing for an unusually long period," Kepler wrote, "snow finally fell on 1 March [1619], and such a large amount that nature appeared to expel in a single impetus the entire amount collected during that absence of moisture."[177] The snow would not stop for another two months, delaying the start of spring until early May. As Kepler recounted, what nature had first appeared to defraud on "it now repaid with interest."[178]

Among the other meteorological responses of the earth, Kepler turned to earthquakes. As one example, he considered the earthquake that had followed the comets of 1618 and produced a landslide covering the alpine town of Piuro.[179] "In the blink of an eye," Kepler wrote, the entire city had met with "sudden disaster" beneath the rock and rubble of a nearby mountain.[180] This was just one example among many of a larger influence that could be brought on by dry conditions. In the event of an earthquake, these exhalations would grow in quantity until "the force of the dry and sulfurous matter beneath the earth" would create an opening in the surface and burst forward.[181] Kepler claimed that the earth could also respond to a comet by giving off a greater exhalation of vapor. In the case of wet weather, the earth could exhale enough vapors to cause flooding and even contaminate the air. "Excessive humidity" and the pollution of the air, Kepler wrote, would then serve as the source of many medical problems:

> And so when this faculty of the earth is stirred by the extraordinary appearance of a comet, it exudes a great many vapors in one place on the earth's surface according to the nature of that part of its body. This is the source of long lasting rain and inundation. And since living beings sustain life by breathing in the air, when there is either excessive humidity or other pollutants resulting from mixtures with nitrous, arsenic, or sulfurous exhalations, there are epidemic illnesses, headaches, dizziness, catarrh, as in the year 1582, and even pestilence, as in the year 1596.[182]

---

[177] Ibid., 252.1–4.

[178] Ibid., 252.5–6.

[179] Ibid., 251.31–33.

[180] *JKGW*, 11,2, 184.1–10. Cf. Schuster and Highland, 2007, p. 3.

[181] *JKGW*, 8, 230.22–23.

[182] Ibid., 230.12–19: "Hac itaque facultate Telluris insolenti Cometae apparitione consternata, uno terrestris superficiei loco multum exsudat vaporum, pro qualitate illius partis sui corporis. Hinc diuturnae pluviae et eluviones. Et quia ex aeris haustu vitam sustentant animalia, una ex aeris seu humiditate nimia, seu inquinationibus aliis per mixtas exhalationes nitrosas, arsenicales, sulphuratas, existunt morbi Epidemici, Cephalalgiae, Vertigines, Catarrhi, ut anno 1582. denique Pestis, ut anno 1596." Cf. *JKGW*, 4, 62.19–24.

Material in essence along with the air and the ether beyond it, the vaporous outflow of the earth participated in the same process of change that Kepler assigned to a natural faculty. This outflow supplied the sea with a regular source of water and allowed for the spontaneous generation of living beings in the same way the ether brought about the birth of comets. "Just as waters, especially salty ones, provide an origin for fishes," Kepler wrote, "so the ether provides an origin for comets; and just as fishes wander upon the waves, so comets wander upon the ether."[183] In response to criticism from Robert Fludd (1574–1637), Kepler would compare this natural faculty of the earth to "that of an animal," as it advanced "into the surrounding air" like "vapors given off by the body."[184] Kepler could make this analogy, he claimed, because the faculty responsible for the spontaneous generation of so many living things on earth acted in the same way as the human body, whose own stock of spontaneous life forms included "lice and similar things."[185] Kepler viewed this faculty as a natural property of many earthly products and an ever-present ability to promote new forms of life throughout the sublunar sphere. Essential to the creation and continuance of life on earth, it played an equally important role in Kepler's greater conception of change in the surrounding heavens.

[183] *JKGW*, 8, 225.2–4; cf. *JKGW*, 4, 59.2–10.
[184] *JKGW*, 6, 408.23–26.
[185] Ibid., 408.23–30.

CHAPTER FIVE

KEPLER'S APOLOGY

While Kepler attributed novelties in the heavens to a natural faculty acting similarly in the sublunar sphere, he could not fully determine the celestial substance on which that faculty acted. "Since no one could be present in the heavens," Kepler wrote, any question concerning their physical nature would rely on "the light rays of the stars" and the perception of their optical properties.[1] Did exhalations of the earth furnish comets and new stars with their physical origins? Or did dense areas of the ether afford the natural faculty with the fuel for these new forms? Either way, Kepler claimed that comets and new stars affirmed the mutable nature of the celestial and sublunar spheres. This claim would meet resistance from one of Kepler's most avid readers, Johann Georg Brengger, a physician in Kaufbeuren who suggested that the pure and simple substance of the ether could not be compared with the elemental world of change. "You learned that the account [*ratio*] of our earth is far different from that of the stars," Brengger wrote to Kepler, "since the bodies of the stars are simple and, according to Aristotle, perfectly pure."[2] The earth, on the other hand, witnessed an endless mixture of bodies that varied in proportion and substance. Deploying ancient reasoning to reckon the heavens and earth differently, Brengger argued that the uniformity of the heavens contradicted their conception as creative in any way. On earth, a natural faculty fulfilled the function of converting rain and snow into a form of "moistening and making the earth fertile, while cleansing the air," Brengger wrote, all of which made life possible for plants and animals.[3] There was no reason for such a function among the stars, Brengger said, "except if one establishes along with Giordano Bruno as many worlds as there are worldly globes."[4] "Unless it is shown either that the stars generate or are nourished or are inhabited by living beings that require nourishment,"

[1] *JKGW*, 1, 245.32–34. Cf. Boner, 2009b, p. 382.
[2] *JKGW*, 16, no. 441, 205–207.
[3] Ibid., 220–224.
[4] Ibid., 224–226.

Brengger claimed, the question of decay and renewal in the heavens was "labored over in vain."[5]

Without determining the material substance of the ether, Kepler explored the possibility that it served as a place for exhalations emitted by the globes of the stars and the planets, including the earth. For his own part, Brengger echoed Kepler by rejecting the "popular opinion of the peripatetics concerning the material of comets taken from earthly exhalations."[6] In a letter to Brengger, Kepler argued that if this opinion were true, the exhalations of the earth would eventually exhaust their source. "For what if the earth does exhale into the ether?"[7] Kepler asked. Counting only the ashes and other things brought up into the air by fire, Kepler claimed that this form of exhalation would spell the end of the earth. "Either say how this [material] returns to the earth," Kepler wrote to Brengger, "or, if you cannot, accept that the earth grows smaller in a single century."[8] Earlier, Brengger had argued that the exhalations stayed within the sublunar sphere and that, by analogy, the exhalations of every planet remained in their own atmosphere. Brengger explained this in a review of Kepler's report on the comet of 1607, which he sent to Kepler in the spring of 1608:

> For just as earthly exhalations reside around the globe of the earth and adhere to it in a certain way without transcending the elemental region, so it is probable that each of the exhalations of the planets (if we may concede to them exhalations) remains around its own globe and is in no way removed from it or diffused into the expanse of the ether.[9]

Brengger would go on to warn Kepler that the suggestion of "a new spirit or motive intelligence," created by God for the purpose of "guiding comets" and expiring as soon as their "task was complete," would offend many, "especially theologians."[10] Kepler explained in reply how Brengger had been right. Theologians had intervened to prevent his report from being published in Leipzig, he wrote, "all on account of that one passage

[5] Ibid., 230–233.

[6] *JKGW*, 16, no. 480, 246–248.

[7] *JKGW*, 16, no. 488, 353–354.

[8] Ibid., 357–359.

[9] Ibid., 260–265: "Imò quemadmodum exhalationes terrenae circa globum terrae resident, eique quodammodo adhaerent ut elementarem regionem non transcendant: ita exhalationes planetarum (si que illis concedantur) quasque circa globos suos haerere, et nequaquàm ab iis avelli aut in coeli spacium diffundi verisimile est."

[10] *JKGW*, 16, no. 480, 260–263.

concerning the creation of spirits."[11] Although he would later revise "the rigidity [*rigor*]" of this passage in Latin,[12] Kepler explained to Brengger that it was more a matter of elaborating on his controversial view than altering it. Neither the matter nor the form of a comet was created from nothing, he claimed, but rather came from ingredients already in existence. Every comet and creature, Kepler wrote, was the composite of "a highly subtle ethereal body" and a spiritual principle that he compared to a ray of light given off by God:

> However, my opinion is not quite so absurd. In every creature, even a spirit, I observe two things, one that I liken to matter and reckon as a highly subtle ethereal body, the other that I liken to form and reckon as a ray of the divine will, to speak symbolically.[13]

By comparing the spiritual principle to "a ray of the divine will," Kepler suggested that whatever this ray fell upon was given "life and reason [*ratio*]" according to the disposition of that particular object.[14] "In the same way," Kepler wrote, "a ray from the sun brings green color to green things, red [color] to red things," and the color of every object according to their own disposition.[15] This was not creation "from nothing [*ex nihilo*]," Kepler claimed, but an act that saw the cometary spirit "produced from the material of the heavens" and "illuminated and informed by a ray of the divine will."[16] It has been argued that Kepler would later replace this claim with the concept of a natural faculty, a more scientific abstraction that would further remove Kepler from his animistic past.[17] Nevertheless, Kepler saw this faculty as similar to the one acting on earth, where he would continue to associate it closely with a sublunar soul. There is no telling where he drew the line comparing the celestial and sublunar spheres.

Kepler's conception of a natural faculty as the perpetual embodiment of archetypal principles suggested that nature and special providence served as part of the same source of origin. When Kepler turned to the

---

[11] *JKGW*, 16, no. 488, 390–391. Cf. Boner, 2012b.
[12] *JKGW*, 16, no. 488, 392.
[13] Ibid., 393–397: "At non valde absurda est mea sententia. In omni creaturâ, etiam in spiritu duo specto, aliquid quod est instar materiae, quod puto esse subtilissimum corpus aetherium, aliquid quod est instar formae, quod puto esse radium vultus divini, ut loquar Symbolice."
[14] Ibid., 398–400.
[15] Ibid., 396–397.
[16] Ibid., 404–405.
[17] Hübner, 1975, p. 244.

new star of 1604, he saw it as the miraculous making of God while at the same time the consequence of comprehensible causes. In a similar vain, Brengger suggested that God made use of mainly "natural means" after the original act of creation.[18] "From the Sacred Scriptures," Brengger wrote, "it appears that God is very rarely found effecting miracles after the creation of the world."[19] For that reason, Brengger accepted the opinion of those who claimed the new star had not been created "anew from nothing [*de novo ex nihilo*]," but rather "from pre-existing material" in the Milky Way.[20] When Kepler turned to comets, the material substance of the ether would serve as an instrument by which a natural faculty produced a comet according to these same principles. While the appearance of a comet or new star from the perspective of the earth required a rare set of circumstances, Kepler assigned their origin entirely to natural causes.

On earth, Kepler identified a natural faculty in the sublunar soul whose sphere of activity involved the expression of new forms. In his *Apology for* The Harmony of the World (1622), written in response to criticism from the Rosicrucian physician and philosopher Robert Fludd, Kepler identified this faculty with an animate motor that acted on material substances. As part of the soul of the earth, Kepler associated this faculty with those works that were "observed in the globe of the earth" that he refused to attribute to "the motions of the elements or the affections of matter alone."[21] In defense of this view, Kepler also condemned Fludd's claim that the air acted as "an inferior spirit of the soul of the world."[22] The air was a material instrument, Kepler argued, and not an animate motor. As proof of this, Kepler pointed to the example of fishes, whose motion was made possible by the preservation of air "in a double bladder."[23] "Even when the soul departs and the lifeless body [of the fish] now suspends from the belly that contains the bladder," Kepler wrote, "this air remains enclosed in that bladder."[24] Kepler extended his argument to other forms of life as well. Although air was essential to the life of every land animal, it was not "the soul itself," Kepler said, and it certainly could not inspire life by simply being inhaled.[25] Rather, "air ventilates the flame of life burning

[18] *JKGW*, 16, no. 441, 265.
[19] Ibid., 263–265.
[20] Ibid., 267.
[21] *JKGW*, 6, 375.17–19.
[22] Ibid., 407.30–31.
[23] Ibid., 408.40.
[24] Ibid., 409.2–4.
[25] Ibid., 409.7–8.

in the recesses of the heart," Kepler wrote, "and prevents the flame from suffocating itself by its own soot."[26] Kepler claimed that the same could not be said when it came to plants. In their case, the air was not thought to preserve "any part of life, not even as an instrument."[27]

Although Kepler acknowledged the part the air played in the preservation of life, he defined it as instrumental and inferior to the authorial role of the soul. In addition to reproaching Fludd for failing to employ "the certainty of mathematical demonstrations,"[28] Kepler criticized him for confounding the motive capacity of the soul with the material by which it operated. That the cause of certain conditions of the weather lay in "a hidden virtue" in the air, Kepler wrote, could only be the opinion of those who had never made an effort "in the investigation of the causes."[29] The air served merely as a material medium, manipulated by the various faculties of the sublunar soul according to a routine that could change in response to the configurations of the heavens. "On earth and in the surrounding air," Kepler wrote, "commotion follows an aspect," and the aspects thus served as "the cause of stirring up storms."[30] As the accidental product of the planetary rays and "a thing of reason" on account of their geometrical harmony, the aspects required the presence of "an animate faculty" in the earth that could recognize and respond to their appearance.[31]

Situating the Soul of the Earth: Elemental Instruments and their Animate Impetus

It was not Fludd who had initiated his dispute with Kepler. In his appendix to *The Harmony of the World,* Kepler gave a critical response to the cosmic picture Fludd had painted in his *History of the Two Cosmoses* (1617). It has been said that Kepler wrote his review primarily to point out where he agreed with Fludd "and where they differed,"[32] and this is expressly what Kepler said he would do. Although he mainly dealt with their differences politely, Kepler could not hide the fact that he believed Fludd had

26 Ibid., 409.8–10.
27 Ibid., 409.11–13.
28 Ibid., 377.8–9.
29 Ibid., 406.16–19.
30 Ibid., 406.28–30.
31 Ibid., 406.28–37. On the intersection of mathematics and meteorology in Kepler's causal conception of sublunar nature, see Magruder, 2006, pp. 236–238.
32 Field, 1988, p. 183.

based his harmonic theory on "his own picture of the world" rather than "the world itself."[33] Kepler referred to Fludd's negligence of mathematics and natural philosophy as a reflection of his ignorance of natural order more generally. Kepler also criticized Fludd for making use of enigmatic images while taking over the opinions of others without examining them in the light of mathematical demonstration. Such a negligence, Kepler said, situated Fludd squarely in the camp of "the chymists, Hermeticists, and Paracelsians," far from the place of precision enjoyed by the mathematicians.[34] The result was a world view composed of confused and false things, Kepler claimed, which stood in contrast to his own approach that took him straight to the heart of nature:

> What Fludd adopts from the ancients, I elicit from the nature of things and establish from their very foundations. Those things that he accepts are confused (on account of the varying opinions of the authors) and he takes over false things, while I advance in the natural order so that everything may be emended according to the laws of nature and confusion avoided.[35]

Much to the aversion of Kepler, Fludd envisioned a series of numerical symbols to associate and elucidate the three parts of his cosmic picture. Kepler condemned these symbols as entirely subjective and the vision of a "numerical prophet" bound in no way by archetypal principles.[36] "Those harmonies that he sets out to show," Kepler wrote, "are mere symbolisms, more poetic or oratorical than philosophical or mathematical."[37] Again, Kepler claimed that Fludd had taken over this view of harmony from the ancients, "who believed that the power [*vis*] of harmonies came from abstract numbers."[38] "It is enough for Fludd," Kepler said, "if he should understand in some numerical way those parts among which a harmony exists."[39] Time after time, Kepler told Fludd he had lost sight of the causal foundation of things. Even when it came to the study of musical instruments, Kepler claimed that he differed from Fludd in the same way "as a

---

[33] *JKGW*, 6, 377.1–3. Cf. Westman, 1984, p. 206.

[34] *JKGW*, 6, 374.20–22.

[35] Ibid., 374.24–28: "... quae ipse transsumit à veteribus; ego è rerum Natura eruo et ab ipsis fundamentis constituo: ipse, quae accepit, ea confusa (propter variantes tradentium sententias), et incorrecta, usurpat; ego naturali ordine procedo, ut omnia secundum leges Naturae sint emendata, et confusio vitetur."

[36] *JKGW*, 18, no. 974, 285–286. Cf. Caspar, 1993, p. 292.

[37] *JKGW*, 6, 374.37–39.

[38] Ibid., 375.35–36.

[39] Ibid., 375.27–28.

practitioner [does] from a theoretician." Fludd had failed to prepare any form of mathematical demonstration in his analysis.[40]

Fludd determined the dimensions of the cosmos according to the mathematical properties of two intersecting pyramids that represented the macrocosm and microcosm. These pyramids expressed a geocentric view that Kepler found fully perplexing. The cosmic harmonies they created, Kepler claimed, were the product of a world picture that Fludd carried about "privately in his own mind."[41] By contrast, Kepler had worked with the actual motions of the planets as they arrived at apogee and perigee and determined their accuracy through astronomical calculation. "I have demonstrated that the full body of harmonic attunement, with all of its parts, is found in the extreme motions of the planets," Kepler wrote, "according to certain measures demonstrated from astronomy."[42] Kepler wished to distinguish his harmonic theory in this way from the abstract ideas of Fludd, who divided the world into three equal parts, "the elemental, the ethereal, and the Empyrean."[43] Despite their many differences, however, it is important to note that Kepler did not disagree with everything Fludd had to say.[44] The determination of cosmic dimensions according to the geometrical properties of two pyramids recalls Kepler's own deployment of geometry in his polyhedral hypothesis.[45] More concretely, Kepler suggested a degree of similarity between their views of the earth as an animate being.

Despite arriving at it "by a very different cause," Kepler believed that he shared his view of an animate sublunar sphere with Fludd.[46] In *The Harmony of the World*, Kepler had argued that the earth exhibited signs of a living, breathing being whose range of activity surpassed the explanatory resources of material existence alone. "If someone should argue that the earth accommodates its respiration, as it were, to the motion of the sun and the moon," Kepler wrote, "he should not be heard with unfavorable ears in philosophy."[47] Kepler suggested that this claim would be especially

---

40 Ibid., 374.12–16.

41 Ibid., 376.36–38. Cf. Schwaetzer, 1997, pp. 80–81.

42 *JKGW*, 6, 376.39–377.1.

43 Ibid., 375.32–36.

44 For some of the similarities between the mathematical cosmologies of Fludd and Kepler, see Huffman, 1988, pp. 55–56.

45 On the two "archetypal scripts" that Fludd and Kepler saw play out in the world, see Westman, 1984, pp. 201–203.

46 *JKGW*, 6, 375.13–14.

47 Ibid., 270.23–27.

persuasive if "some evidence of flexible parts" were found in the depths of the earth that allowed for the respiration of "lungs or gills."[48] Rather than regard it as a flight of fancy, Kepler argued in his apology that Fludd's outright rejection of this idea only further revealed his inadequacy in mathematics. "When you reject the respiration or reciprocation of the earth as the cause of the ebb and flow of the sea," Kepler wrote, "you reveal that you are not a geometer."[49] Judging the depth of the ocean to reach roughly 900 German miles, Kepler claimed that the smallest pattern of contraction and expansion at a point along the ocean floor would produce the ebb and flow found on the surface. Even if this meant that the interior of the earth was "full of water," Kepler believed that a form of contraction or inhalation measuring only a few feet would be more than enough.[50]

In response to Kepler's critical review, Fludd fired back with a tract of no fewer than fifty-four folios, *The Stage of Truth* (1621), which he published in Frankfurt. Fludd's reply reveals a reluctance to consider (or even contemplate) the mathematical side of Kepler's argument and a far greater concern with their conflict in the arena of cosmology.[51] Kepler countered with a detailed defense that he published in the same city a year later. In his *Apology for* The Harmony of the World (1622), Kepler reproduced line after line from Fludd's reply with the aim of refuting his rival while clarifying his own convictions. As a result, Kepler shed further light on his cosmology and brought greater clarity to his own account of the soul of the earth.

In a letter to the philologist Matthias Bernegger (1582–1640), Kepler described his *Apology* as an undesirable task. It was "a response to the most incompetent book by Robert Fludd," Kepler wrote, which he had promised reluctantly to Gottfried Tambach (1607–1632), his publisher in Frankfurt.[52] Despite his distaste for the work, Kepler composed many of his arguments carefully and courteously. Playing on the theatrical title of Fludd's text, Kepler introduced his *Apology* as an offstage exposition where impartial honesty held sway over "empty ambition."[53] Impressed

[48] Ibid., 270.27–29. Kepler's theory of earthly lungs was later rejected by Libert Froidmont (1587–1653), along with the idea that the earth was an animal. See Martin, 2011, p. 101.

[49] *JKGW*, 6, 418.33–35.

[50] Ibid., 418.37–39.

[51] On Fludd's failure "to follow much of Kepler's mathematical argument," see Field, 1988, p. 184.

[52] *JKGW*, 18, no. 919, 9–11.

[53] *JKGW*, 6, 383.29–30.

by "the customary benevolence" that Kepler expressed towards Fludd,[54] Philip Müller (1583–1659) praised the *Apology* as a testament to intellectual tolerance. "How humanely and politely has Robert [Fludd] been held by you," Müller declared, "in both your appendix and the *Apology*!"[55] Fludd was "a stubborn man," Müller wrote, who knew nothing—unless, of course, he knew that he knew nothing.[56] A professor at the University of Leipzig, Müller was struck above all by Kepler's capacity to consider the criticism of others, "even the ignorant," in such a careful way.[57]

Kepler said he had not written the *Apology* to cause further confrontation. He objected to the claim that necessity had compelled Fludd to respond to Kepler's comparison of their work. Nowhere, Kepler wrote, had he "ever raised anything against" Fludd in his appendix.[58] On the contrary, Kepler had discussed his theory of world harmony in relation to Fludd's relevant views in order to distinguish them and thereby "instruct [his] readers."[59] As a matter of fact, Kepler's appendix was as much about Ptolemy as it was about Fludd. At one point, Kepler accused Ptolemy of the same crime of committing to "symbolisms" that were "neither causal nor natural, but poetic and oratorical."[60] "There is no difference between you and Ptolemy," Kepler quipped in the *Apology*, "except that you are still alive and he preceded us by fifteen centuries."[61] Rather than adopt the confrontational style of his counterpart, however, Kepler would also concede several things to Fludd. And he expressed his sincere approval of Fludd's resolution to consider their work comparatively, "leaving aside acrimony."[62] Whether or not Fludd actually fulfilled this promise, Kepler would still remark on those points worth praising in *The Stage of Truth*.

Fludd often made reference to the Aristotelian principle that the substance of the earth was "best suited to idleness and contrary to the activity

---

54 *JKGW*, 18, no. 936, 8.

55 Ibid., 11–12.

56 Ibid., 14–17. Müller would later claim, however, that his inexperience and youth put him in a position of ignorance that he shared with Fludd and others. "Nevertheless, there is something that I share with them," Müller wrote, "from which I fear more and more. And what is it? Ignorance, inexperience, and youthfulness in those things that I profess but do not understand." See ibid., 40–45.

57 Ibid., 10.

58 *JKGW*, 6, 383.18–19.

59 Ibid., 384.12–16.

60 Ibid., 371.3–5.

61 Ibid., 384.28–30.

62 Ibid., 384.8–10.

of a soul."[63] Fludd believed that the body made up of this material and positioned motionless at the center of the cosmos could not contain a soul. He suggested that a soul inhabited the air rather than the terrestrial interior. In his reply, Kepler compared the earth to the human body, which was claimed to consist mainly of earthy matter. "The earth that is in my body," Kepler wrote, "is best suited to idleness," but this did not mean that a soul did not dwell in his body. Although the body required rest, this was often brought on, Kepler claimed, by an "agitation of mind" as he sat busy at work under the strain of his soul's activity:

> And yet the earth that is in my body is best suited to idleness and contrary to the activity of my soul. For while I write with such agitation of mind, my eyes close, my head descends, my pen must be laid down, and my body, finally yielding to this earth, must be sent to bed. This must come between that agitation of mind, so that the body may enjoy idleness as well as rest. It does not then follow, however, that there is not a soul in this earthly body.[64]

Arguing from the above analogy that a soul occupied the earth, Kepler would later employ another analogy to explain the earth's diurnal rotation. In the same way that children spun a top, Kepler said in the *Epitome of Copernican Astronomy*, God could have set the earth spinning with a single turn. Kepler claimed that "every subsequent rotation" would then have repeated this original impression [*impressio*], which extended "with unbroken vigor down to this day."[65] Since then, the earth had reportedly revolved a total of 2,000,000 times without any obstruction from friction, the density of the ether, or the "internal heaviness [*gravitas interna*]" of the earth.[66] To fulfill this uniform motion, Kepler speculated that the internal fibers of the earth were woven circularly around a central axis by "a corporeal faculty."[67] Formed in this way, the physical structure of the earth would ensure that it continued to move flawlessly.

---

[63] Ibid., 415.36.

[64] Ibid., 415.37–416.3: "Atqui et hoc terrae quod inest in corpore meo, quieti est aptissimum, estque contrarium actui animae meae: dum enim agitatione mentis ista scribo, connivent oculi, defluit caput, denique obsecundandum huic terrae, dimittendus calamus, corpus in torum abjiciendum; intercedendum huic agitationi mentis, ut quiete et somno et corpus fruatur. Neque tamen inde sequitur, non inesse in hoc terreno corpore Animam."

[65] *JKGW*, 7, 89.29–32.

[66] Ibid., 89.31–34.

[67] Ibid., 89.43.

Kepler's theory of the axial rotation of the earth thus began with an original impression that "God the Creator incited in the globe of the earth."[68] This impression had been "transformed or coalesced into a corporeal faculty," Kepler claimed, which in turn had created a series of straight fibers extending latitudinally in the same direction. If the Earth were sliced equatorially, Kepler suggested, the interior on either side would reveal those rectilinear fibers situated circularly around the central axis like the growth rings of a severed tree. Configured in the same way as "the mobile body of a spinning top,"[69] they would forever limit the slightest loss of motion. That the straight-line structure of these fibers also produced a magnetic field that made the earth approach and recede from the sun at different points—straying, that is, from a perfectly circular course and tracing the form of an ellipse—was not perceived by Kepler as a problem. Although the form of a body furnished it with "a cause of motion in a certain direction,"[70] Kepler said, it often happened that multiple forms existed in a single body that could perform multiple motions. "In the same way that a magnet attracts iron at one end and repels it at the other,"[71] Kepler explained, the subterranean structure of the earth allowed for axial rotation as well as an elliptical course. Kepler turned again to the human body to give an example of something whose fabric of fibers also allowed for a variety of functions. The stomach, he suggested, had been found by physicians to consist of "three series of fibers," with each series accounting for a different form of motion and a corresponding faculty:

> The physicians find in the substance of the stomach an example of not only a double entwining, like that [of the earth], but clearly a triple entwining. In the same way, they distribute among those three series of intertwined fibers three faculties of the stomach, the attractive, the retentive, and the expulsive.[72]

Kepler's analogy of the human body did not end with the fibrous structure of the stomach. He also argued that the earth's internal fibers served as instruments in the expression of a motive impression in the same way

---

68 Ibid., 90.13–14.

69 Ibid., 90.5–6.

70 Ibid., 90.34–35.

71 Ibid., 90.35–38.

72 Ibid., 90.29–32: "Exemplum huius fibrarum implexionis non geminae tantum ut hic, sed plane triplicis, habent Medici in substantia ventriculi, qui consimiliter inter tres illos fibrarum ordines, implexos mutuo, tres ventriculi facultates distribuunt, attracticem, retentricem, expultricem."

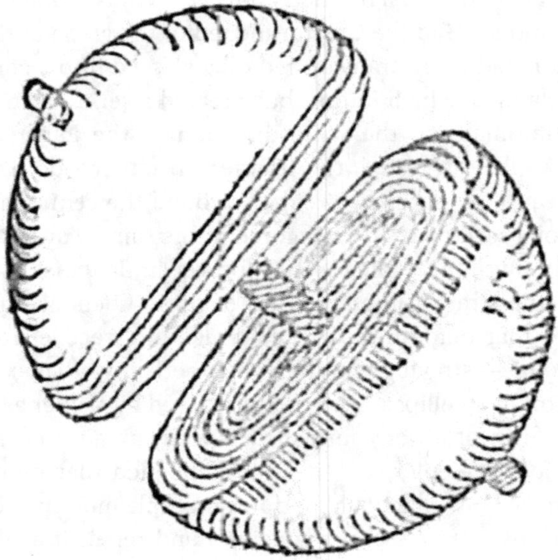

Figure 4. Hemispherical section revealing the latitudinal structure of the earth's interior. Johannes Kepler, *Epitomes astronomiae copernicanae liber I* (Linz, 1618), p. 117. Courtesy of the Dibner Library of the History of Science and Technology, Smithsonian Library Institutions.

that the various parts of the human body acted out the impetus of a soul. Although "the bones, joints, ligaments, muscles, and nerves in the human body [were] perfectly suited to motion," they were not "the primary motive cause" for Kepler, but "the instruments of a soul for moving the body."[73] In the earth as well as the human body, a set of material instruments were necessary to carry out the command of a soul. The initial source of the earth's rotation was supplied by God, "the Creator and prime mover," while the continuation of it recalled the original form of motion.[74] According to this account, Kepler claimed that "every soul and spirit" bore the mark of God, the eternal source of generation, in the same way that a shadow bore the form of a body.[75]

Kepler believed that the sun also played an important part in the rotation of the earth. That the earth rotated 365 $^1/_4$ times a year, Kepler

[73] Ibid., 91.10–13.
[74] Ibid., 91.32–35.
[75] Ibid.

Figure 5. Hemispherical section revealing the longitudinal structure of the earth's interior. Johannes Kepler, *Epitomes astronomiae copernicanae liber I* (Linz, 1618), p. 121. Courtesy of the Dibner Library of the History of Science and Technology, Smithsonian Library Institutions.

argued, was a product of "the perpetual presence of the sun," which also made the earth spin 5 $^1/_4$ times faster than the archetypal number of 360.[76] If "the inner motive force [*vis*] of the earth were not invigorated by the sun," Kepler wrote, "the earth would advance somewhat more slowly on its axis," reaching "only 360 revolutions in the same annual interval."[77] In addition to this influence, the sun caused other inconsistencies, such as the slightly slower rotation of the earth in the summer than in the winter. This had been first brought to light by Tycho and then converted "into physical form" by Kepler,[78] who determined that the greater distance of the earth during the summer in the northern hemisphere resulted in a decrease in the diurnal motion. Kepler denied that any such irregularity

[76] Ibid., 316.32–33.
[77] Ibid., 316.33–35.
[78] Ibid., 316.37–42.

resulted from "an innate principle [*insitum principium*] of the earth," since this sort of principle would only act uniformly.[79]

How exactly could the distance of the earth from the sun cause a corresponding difference in the diurnal motion? Kepler claimed it was "highly probable" that "the motive virtue [*virtus motoria*] of the earth" was moved in varying degrees by the illumination of the sun.[80] This provided a physical explanation for why the earth spun slightly slower during the northern summer than during the southern one. "Since the physical equation of time requires unequal diurnal revolutions in the same way the distance of the earth varies from the sun," Kepler wrote, the amount of light that arrived at the earth would be "denser at a shorter distance" and feebler at a farther one.[81] Here was a direct proportion between the rotation of the earth and the intensity of light that the earth received. In this way, Kepler wrote, "the amount of light there is at any moment may be suited to the measure of distance for determining this acceleration over the year."[82]

Kepler gave his account of the diurnal motion in combination with what he called "a magnetic affection [*affectio magnetica*] in the earth."[83] According to this source of magnetic sympathy, the moon orbited the earth while the earth orbited the sun. Kepler suggested that while the solar rays accelerated the diurnal motion, this magnetic affection caused the earth to approach or recede from the sun depending on the distance and the inclination of the earth's axis.[84] Profoundly impressed by the magnetic philosophy of William Gilbert, Kepler claimed that Gilbert had established the existence of "a magnetic faculty in the globe of the earth."[85] In a letter to a patron, Kepler marveled at Gilbert's ability to make use of observational evidence to such great effect that he had "closed the mouth of every critic."[86] It was Gilbert, Kepler wrote, who had first claimed that "the fibers or magnetic filaments of the earth extend in straight lines from South to North."[87] And as the earth was thought to possess "a great affinity with the other planets," it shared with them "a magnetic disposition

---

79 Ibid., 316.42–317.1.

80 Ibid., 317.41–44.

81 Ibid., 317.44–318.3.

82 Ibid., 318.5–6.

83 Ibid., 319.43–44.

84 Kepler's conception of this magnetic affection has often been mistaken for a set of purely mechanical principles. See, for example, Aiton, 1975, pp. 66–68.

85 *JKGW*, 16, no. 441, 236–241.

86 *JKGW*, 14, no. 242, 212–213.

87 *JKGW*, 7, 334.4–5.

[*dispositio magnetica*]" that caused their deviation from a purely circular path around the sun.[88] As Kepler wrote in a letter to Christopher Heydon (1561–1623), the notion of a magnetic faculty lay at the heart of a new and revolutionary view of planetary motion. Such was the grand vision of the *New Astronomy* (1609), whose account of the eccentric orbit of Mars Kepler applied to the path of every other planet:

> So, what is it that carries the planets around the sun? For Tycho and Copernicus agree that [they move around it]. What, then, if not a magnetic effluvium from the sun? And what is it that makes the planets eccentric from the sun and forces them to approach the sun and recede from it? Why, a magnetic effluvium from the very bodies of the planets, and the direction of [their] axis. And all of those calculations on Mars have been compared in such a way that either it is necessary that they be false or common to every planet with respect to their quality.[89]

Fludd fathomed another form of interaction with the sun, a spirit that extended from it and gave "life and soul" to the planets.[90] Kepler criticized this conception for attributing the agency of the astrological aspects to the planets rather than to the soul of the earth. "To every moment," Kepler wrote, "you assign a new soul of the planets from the spirit of the sun," and so "two planets from their places in the heavens may stir up the weather in the sublunar world."[91] The earth was responsible for acting, Kepler argued, not the planets, which merely served as a source of influence on natural processes already underway. That the planets were directly responsible for the meteorological activity of the earth was also opposed by the occurrence of different conditions at the time of a particular configuration in the heavens. Different areas of the earth witnessed different weather under the appearance of a single aspect, and the recurrence of a particular configuration did not always result in the same regional responses. "How much more probable is it," Kepler asked Fludd, "if I put that perpetual spirit in sublunar nature?"[92]

---

88 *JKGW*, 16, no. 441, 249.

89 *JKGW*, 15, no. 357, 49–58: "Nam quid est, quod Planetas circa Solem rapit? Consentiunt enim Tycho et Copernicus in eo. Quid enim nisi effluvium Solis magneticum? Quid vero est, quod Planetas facit à Sole eccentricos, quod cogat ipsos ad Solem accedere, ab eo recedere? Nempe effluvium ex ipsis Planetarum corporibus magneticum, et directio axis. Atque haec omnia ratiocinia in Marte sic sunt comparata, ut aut falsa esse necesse sit, aut omnibus Planetis, quoad qualitatem, communia."

90 *JKGW*, 6, 407.24.

91 Ibid., 407.25–27.

92 Ibid., 407.28–30.

In fact, the variability of the weather further convinced Kepler of a sublunar soul. He interpreted this diversity as the expression of an animate being whose instinctual response to the heavens conformed to the natural conditions prevailing at any one point along the surface. And just as the configuration of the heavens could intensify meteorological activity, Kepler claimed that the earth could also stir up storms in response to stones thrown into valleys and other openings. Kepler had expressed doubt about this idea in a letter of 1602 to David Fabricius, where he recalled his ascent of Mount Schökl on a summer's day in Styria. "When we ascended the mountain," Kepler wrote, "the air was clear and hot, and there was a great thunder storm in Graz two miles away."[93] Rumor had it that a small stone thrown into a hollow of this mountain could start a hailstorm. "I reckon this as a deception of rumor," Kepler said, "for we did not find a hollow and there was still a hailstorm, unless perhaps we had irritated the mountain on our ascent, of course."[94] Kepler would later reverse his position and argue in the *Apology* that storms could arise "from the tossing of small stones into the openings of the earth."[95]

Kepler considered other conditions, such as fog and rain, as further signs of a vibrant being whose flame of life burned brightly in the earth. While the earth's daily rotation was regarded as "an ordinary and everlasting labor," Kepler referred to rain as "an extraordinary effervescence" that was often brought on by the appearance of an astrological aspect.[96] Kepler argued that the motion of the earth occurred apart from these outpourings in the same way as "a camel may proceed on its path without interrupting its steps, while all the while now this part, now that part of its hide corrugates to kill flies."[97] Rather than rely on the heavens for ordinary activity, the earth made use of the aspects strictly as a special source of stimulus. "It is certainly not my dogma," Kepler wrote, "that motion or vegetation in plants and animals either existed or can exist before or without the motions of the heavens."[98] Kepler referred Fludd to Book 4 of *The Harmony of the World*, where he would find that "sublunar nature

[93] *JKGW*, 14, no. 239, 520–523.
[94] Ibid., 539–541.
[95] *JKGW*, 6, 418.27–28.
[96] Ibid., 418.20–22.
[97] Ibid., 418.22–24.
[98] Ibid., 443.41–444.1.

always attends to its own work while seizing on some special sources of impetus from the stimuli of the aspects."[99]

In words that Kepler reproduced from *The Stage of Truth*, Fludd had suggested that the sublunar soul was "in the air."[100] Naturally situated in the elemental system, the air was thought by Fludd to bring life "to the land and sea" and everything made up of the two.[101] In his pointed reply, Kepler claimed that if Fludd had read Aristotle more carefully he would have found that the air served merely as a material instrument while the heat of the sun supplied the real source of life. "Aristotle attributes all of these things to the heat of the sun," Kepler wrote, while the air acted as "one of its instruments."[102] Kepler argued that the soul of the earth advanced from below the surface into the surrounding air. That the sublunar soul derived from "the motion of the air," Kepler said, was an expression "foreign to my philosophy."[103] Closer to the truth would be "to say that the air is generated by the soul of the earth."[104] The sublunar soul was responsible for rarefying the subterranean humors, Kepler wrote, and this vapor rose above the surface in the same way as the sweat of human beings. This produced a wide variety of vapors that differed in density. When they finally found an outlet, they acted on the weather according to their quantity and the climate into which they were introduced.[105]

Through extensive observation of the weather, Kepler had determined when the soul of the earth was especially prone to produce vapors in response to the appearance of the heavens. In an earlier text, he had claimed there was an immediate form of contact between "light and every creature" that could account for this 'conversation,' so to speak, with the heavens.[106] The light rays from the heavens arrived at the earth instantaneously, where the world of living beings received their stimulus and responded with their own form of motion. "Too coarse for the celestial

99 Ibid., 444.2–4. Cf. Simon, 1979, p. 180: "It is in Chapter 7, Book 4 of *The Harmony of the World* that Kepler develops in the most synthetic manner his conception of the soul in general and that of the earth in particular."

100 *JKGW*, 6, 408.22.

101 Ibid., 408.31.

102 Ibid., 408.32–33.

103 Ibid., 409.17–18.

104 Ibid., 409.18.

105 On Kepler's reluctance to predict the weather without referring to regional differences, see Rabin, 1997, pp. 751–753.

106 *JKGW*, 4, 218.29–30.

stimulus to pass through and take effect," the elements were said to participate in this response only insofar as a soul acted on them:

> The elements are too coarse for the celestial stimulus to pass through and take effect, but the harmony of the rays passes immediately into the souls themselves, and there the impulse first occurs and motion then follows in the bodies, the humors, and the elements.[107]

Kepler suggested that the activity of the earth grew stronger when the harmony of the rays grew greater. And when Brengger challenged this view in the light of the motion and decentralization of the earth in the new cosmology, Kepler argued that these changes made no difference. "You say that those [aspects] that are formed outside the center of the world," Kepler wrote, "are particular and therefore weaker."[108] Yet there was no reason why the aspects required a central point of reference, Kepler replied, since their geometry did not require a specific location. Every point in Kepler's finite view of the cosmos was particular, and there was no telling the aspects apart strictly on the basis of their singularity. "I deny that those [aspects] that join together at the center of the world are universal," Kepler wrote, "for any one aspect exists at a single point in the world while every other one [exists] at another, and so every aspect is particular."[109] When it came to astrology, Kepler continued to center the stars on the earth even after he put it in motion at a point away from the center of the cosmos.

As for those aspects that took shape around a point other than the earth, Kepler offered little speculation. In his response to Galileo's discovery of four 'Medicean Stars' orbiting Jupiter, Kepler had conjectured that the configuration of those satellites could inspire the admiration of their Jovian observers. "For whose good do the four satellites revolve around Jupiter at different distances and times," Kepler asked, "if there is no one on the globe of Jupiter who may observe this remarkable variety with their own eyes?"[110] Kepler made no claim to the astrological significance of the four satellites. He expressed concern about their impact on his system of aspects on earth, however, which relied on a complete and

---

[107] Ibid., 218.31–35: "Und hingegen seynd die Elementa viel zu plumb darzu / daß der himmlische Antrieb durch sie solte zugehen und geschehen / sondern die Harmonia radiorum gehet immediate in ipsas animas, unnd da gechicht alsdann der impulsus, in parte principe, motus origine, so folget alsdann die Bewegung erst in die Lieber Humores und Elementa."

[108] *JKGW*, 16, no. 488, 385–386.

[109] Ibid., 387–389.

[110] *JKGW*, 4, 306.13–15.

accurate astronomy. Aware that his system had been based on an incomplete number of celestial bodies, Kepler suggested that Galileo's discovery did not detract from the influence of those aspects he already accepted. "Shall there be those to whom our doctrine of aspects may appear false," Kepler asked, "since down to the present day we did not know the number of satellites making the aspects?"[111] And yet the earth determined the efficacy of the aspects, Kepler quickly replied, not the other way around. "The stars do not act in us themselves," he explained, "but their aspects become the object and stimulus of terrestrial faculties participating in reason without discourse."[112] In the end, Kepler claimed that his close knowledge of nature and careful observation of the weather had ensured that his astrology would survive the discovery of new satellites without any fundamental damage.

Kepler gave particular emphasis in his *Apology* to the earth's exhalation of vapors and their similarity to human perspiration. Just as a human being produced sweat in response to heat or began "seething from the motion of the body,"[113] the earth produced vapors in greater abundance under the illumination of the sun or during times of great agitation. The nature of the weather that then followed depended in part on the nature of those vapors. More importantly, however, the soul of the earth drew an impulse for this activity by relating the configuration of the heavens to an internal archetypal constitution. As Kepler explained to Fludd, the soul of the earth was, like every other soul, essentially a circle. And from the division of the circle came the regular plane figures, which Kepler identified as the foundational elements of the intellectual harmonies that underlay the aspects. "The circle shines forth in the soul," Kepler declared, and from the circle derived "the divisions according to the regular plane figures."[114] Kepler referred to Proclus as an authority on this view of the soul and admonished Fludd for following Iamblichus while forgetting one of his most famous disciples. Kepler explained that the angular separation of two or more celestial bodies on the circumference of the circular soul of the earth amounted to the degree of difference that those bodies measured along the zodiac. Seen as the center of a circle, the earth then

---

[111] Ibid., 306.21–23.
[112] Ibid., 306.26–28.
[113] *JKGW*, 6, 409.21–22.
[114] Ibid., 404.27–28. Although Kepler is widely known for breaking the 'spell of circularity' by introducing the elliptical orbit, the circle continued to play an essential part in his system of archetypal principles. See Brackenridge, 1982; Samsonow, 1986, pp. 30–35; Illmer, 1991, pp. 96–114.

distinguished from this mosaic of moving bodies particular moments in which their positions on the perimeter matched the angles of the vertices of regular plane figures.

Faced with an endless number of regular plane figures, Kepler accepted as archetypal only those that could be constructed with a compass and ruler. Only these figures were thought to express rational proportions that the human mind could perceive and the human hand could reconstruct. As an animate being that became aware of these proportions instinctually, the soul of the earth could distinguish their momentary manifestation from an otherwise constant cacophony of celestial configurations. Kepler claimed their identification was often accompanied by an acceleration in the normal course of sublunar nature. How the soul of the earth recognized and responded to the appearance of these principles in the heavens, Kepler argued, had a lot to do with harmony. The sublunar soul sensed in the aspects many of the same archetypal principles that underlay the musical consonances. Kepler observed that the triangle, for example, divided the circle into parts that determined "the terms of a certain harmonic proportion."[115] The geometrical principles of these proportions acted on the soul objectively, Kepler said, since they were "a thing of reason." The response they elicited was the subjective responsibility of the sublunar soul, however, whose acceleration of the weather Kepler compared to the dancing of a farmer. Just as the farmer could dance to a melody without fully knowing the mathematics of music, so the earth could move the air in response to the geometry it sensed in the aspects instinctually. Kepler expressed this analogy in his account of how the influence of the heavens lay with "some animate faculty" in the earth:

> Yet geometry is a thing of reason that has no efficacy of its own. It is therefore necessary that geometry, lying here in the aspects, act objectively. Yet it cannot act objectively on anything except for the animate faculties, as music does on the hearing of a dancing farmer. And so what causes the motion of the air according to the order of the aspects formed on earth participates in an animate faculty and is present in the earth.[116]

---

[115] *JKGW*, 6, 404.38–405.1. On the presence of archetypal principles in Kepler's conception of the soul and their counterpart in harmonic proportions, see Fabbri, 2003, pp. 43–46.

[116] *JKGW*, 6, 406.32–37: "At Geometria Ens rationis est, quod per se nullam habet efficaciam. Oportet ergo ut Geometria haec latens in aspectibus, agat objectivè. At objectivè in nullam rem agere quicquam potest, nisi in facultates animales, ut Musica in auditum Agricolae saltantis. Quod igitur causatur motus aeris ad praescriptum Aspectuum in terrâ formatorum, id est animalis facultatis particeps, et in terrâ praesens."

Kepler argued that the harmonic proportions he associated with the aspects consisted "not in being, but in becoming."[117] As he had said in *The Harmony of the World,* the soul of the earth rejoiced in those proportions when they appeared in the aspects and made use of them "for fulfilling its own functions."[118] "According to the order of the proportions that occur among the radiations of the stars," Kepler wrote, "the soul of sublunar nature stirs up the weather."[119] And what the soul of the earth did to stir up the weather the human soul did in the way of dance and song. Kepler believed that men moved "their bodies and tongues" according to the very same laws that underlay music as well as the science of the stars.[120] These laws of proportion had been put to use by "God the Creator," Kepler said, "who distributed the motions of the heavens according to the harmonic proportions."[121] And when it came to the aspects, Kepler claimed that the earth witnessed periods of greater activity after configurations expressing these same proportions appeared in the heavens. "Everything is alive when the harmonies endure," he wrote, "and sluggish when they are disrupted."[122]

How exactly did Kepler distinguish the handful of influential configurations from the others? In addition to their mental comprehension and the 'compass and ruler' standard of manual reconstruction, Kepler suggested another property that set the aspects apart: congruence. At the beginning of Book 2, "On the Congruence of Harmonic Figures," Kepler described congruence as the emergence of "the well-springs of the harmonic proportions" in "some bodily form."[123] Viewed in this way, congruence was defined as the divine constitution of bodies according to archetypal principles. "Lying in the most holy mind of God from eternity," these principles were thought to find their physical expression in the property of congruence.[124] Kepler believed it had obliged the archetypes to abandon their abstract state and "break out into the work of creation."[125] The heavens could exemplify these principles by providing "a prelude of effects beyond geometry and the thoughts of the mind on natural and celestial

117 Ibid., 105.16–17.
118 Ibid., 105.18–19.
119 Ibid., 105.22–24.
120 Ibid., 105.30–31.
121 Ibid., 105.21–22.
122 Ibid., 105.32–33.
123 Ibid., 67.19–26.
124 Ibid., 67.28–29.
125 Ibid., 67.30–31. Cf. Hübner, 1975, p. 182.

things."[126] The stimulus of the aspects on the earth, Kepler said, stemmed from the ability of the sublunar soul to discern from their appearance those archetypal principles that were "nothing other than the very essence of God."[127]

As proof that these principles were an essential part of the sublunar sphere, Kepler identified a number of examples of their occurrence in nature. If the soul of the earth could recognize and respond to archetypal principles as they appeared in the heavens, he reasoned, it could also express them in physical form. In this way, the earth was thought to convey archetypal figures in every sublunar phenomenon through the property of congruence. Kepler had argued this in *The Six-Cornered Snowflake,* where he witnessed signs of the archetypal essence of the sublunar soul everywhere. From the growth of leaves in groups of five to the rhombic arrangement of pomegranate seeds, Kepler believed there were countless opportunities "to behold the beauty or singularity of the form that marked [*characterisavit*] the soul of those plants."[128] The geometrical imprint of the Creator had been preserved "down to the present day by the remarkable nature of the animate faculties," Kepler wrote, even in something as insignificant as a snowflake, whose hexagonal form was not without divine cause:

> And after examining everything that occurred to me, I thus feel that the cause of the hexagonal form of the snowflake is nothing other than what is found in the regular figures of plants and the constant numbers. And since nothing occurs in those things without highest reason—not even what may be found by discursive reasoning, but what was originally in the plan of the Creator and preserved from the beginning down to the present day by the remarkable nature of the animate faculties—not even in a snowflake do I believe this regular figure exists accidentally.[129]

Kepler argued that the archetypes were expressed by the animate faculty of every living being in the sublunar sphere. He vowed that he would not

---

[126] *JKGW,* 6, 67.23–25.

[127] Ibid., 55.11–20. On Kepler's view of the divine nature of geometry and the geometrical essence of God, see Kozhamthadam, 1994, p. 20.

[128] *JKGW,* 4, 270.19–21.

[129] Ibid., 275.14–20: "Itaque omnibus examinatis, quae occurrebant, sic ergo sentio, causam figurae in nive sexangulae, non aliam esse, quam quae est figurarum in plantis ordinatarum, numerorumque constantium. Ac cum in his nihil fiat sine ratione summa, non quidem quae discursu ratiocinationis inveniatur, sed quae primitus in creatoris fuerit consilio, et ab eo principio hucusque per mirabilem facultatum animalium Naturam conservetur; ne in Nive quidem hanc ordinatam figuram temerè existere credo."

dignify the chymical doctrine of Fludd, "together with Hermes and Paracelsus, not even with a word."[130]

As the above passage suggests, Kepler believed that the animate faculties continued to express the archetypal principles originally impressed on the sublunar soul. In *The Six-Cornered Snowflake*, Kepler anticipated the objection that these principles were particular and not universal. "Someone may object that for every plant there is a particular animate faculty," he had written, "since every specimen of plant also subsists separately."[131] Despite the separate existence of every species, however, Kepler held firmly to the universal presence of the archetypal principles. They were conveyed from generation to generation by "a universal faculty in the earth," Kepler argued, and stemmed originally from the sublunar soul.[132] This faculty was "in itself one and the same," Kepler said, while it divided itself among the bodies of every living being and shaped them according to their own natural properties.[133] "[It] implants itself in bodies," Kepler wrote, "and designs one and the next according to the internal condition of the material or the external circumstances."[134] In the case of the snowflake, Kepler claimed that this faculty expressed the form of a hexagon "by the contraction of the parts" in vapor when the temperature grew cold.[135]

In his account of harmony in the human soul, Kepler recalled the Proclean expression of anamnesis in the commentary to Euclid's *Elements*. The soul would "reawaken, as if from sleep," when it identified archetypal principles in sensible things.[136] Applied to the soul of the earth, this idea suggested that the aspects served to recall the sublunar soul to the same principles. "To recognize," Kepler wrote, "is to compare an external sensible with internal ideas and judge it congruent with them."[137] Interpreting the words of Proclus in a Christian sense, Kepler viewed the soul as an "exemplar of the Creator" that was in no way "a blank sheet [*tabula rasa*], empty of every ratiocination," but inscribed with the archetypes from the very beginning.[138] Through their momentary expression in the heavens, these principles resonated with a sublunar soul that had been

130 *JKGW*, 6, 399.2–3.
131 *JKGW*, 4, 276.5–6.
132 Ibid., 276.13–14.
133 Ibid., 276.18.
134 Ibid., 276.19–20.
135 Ibid., 276.22.
136 *JKGW*, 6, 226.5–8.
137 Ibid., 226.6–7.
138 Ibid., 220.40–221.4. Cf. Hübner, 1975, p. 234.

carved from the geometrical quarry of the Creator. It is important to note that Kepler did not conceive of these principles as purely numerical entities, but that congruence consisted in "figures as a whole."[139] Kepler believed that archetypal figures could be reconstructed by the inscription of polygons in a circle and that the aspects effectively carried out this reconstructive process. Although Kepler admitted that the number of constructible figures extended to infinity, he claimed that only twelve figures were congruent. In turn, these figures underlay twelve of the aspects that Kepler accepted as influential. As he explained, they measured the arc of the zodiac in the following way: 180°, or opposition, "from the diameter of the circle;" 90°, or quadrature, "from the square;" 120°, or trine, "from the triangle;" 60°, or sextile, "from the hexagon;" 45°, or semiquadrature, "from the octagon;" 135°, or sesquiquadrature, "from the octagonal star;" 30°, or semisextile, "from the dodecagon;" 150°, or quincunx, "from the dodecagonal star;" 72°, or quintile, "from the pentagon;" 108°, or tridecile, "from the decagonal star;" 144°, or biquintile, "from the pentagonal star;" and 36°, or semiquintile, "from the decagon."[140] Kepler also accepted conjunction, which measured 0° of separation.

Fludd accused Kepler of arguing for more aspects than "integral parts in the zodiac."[141] According to Fludd, the twelve houses of the zodiac expressed an equal division of the circle in a way that Kepler's thirteen configurations never could. After explaining why he accepted "not more, but fewer aspects than integral parts,"[142] Kepler turned to Fludd's claim that Kepler had arrived at the aspects by way of the regular polyhedra. "I tried something of this sort in the *Cosmographical Mystery*," Kepler said, but he had since dispelled "those old illusions in *The Harmony of the World*."[143] "Beware, reader," Kepler wrote, "that you entrust to Robert any of those things he attributes to me that my book itself can bear witness."[144]

Configurations and Consonances: The Earthly Orchestra of the Astrological Aspects

Kepler no longer derived the aspects from the regular polyhedra, nor did he claim they shared the same geometrical foundations as the consonances.

---

139 *JKGW*, 6, 108.3.
140 Ibid., 250.31–251.4.
141 Ibid., 412.30–31.
142 Ibid., 412.39–40.
143 Ibid., 413.4–6.
144 Ibid., 413.7–9.

Until 1608, Kepler had thought there were as many aspects as "harmonic sections in music."[145] Now the number of aspects had grown from eight to thirteen, breaking their one-to-one correspondence with the consonances. By opening the door to more aspects, Kepler made it clear that they did not derive from archetypal principles in precisely the same way as the consonances. While he continued to insist that they stemmed from similar foundations, Kepler distinguished the aspects as entirely "within the bounds of the circle."[146] The consonances, by contrast, were ultimately deduced from the comparison of straight lines. As Kepler explained in his *Apology*, this meant that he still derived the octave and the aspect of opposition from "the same geometrical foundations." What Fludd had forgotten when he criticized Kepler for this, he said, was how widely their association had been accepted by astrologers since the time of Ptolemy:

> I deduced the octave and the aspect of opposition from the same geometrical foundations. I am not the original author of this association. The entire race of astrologers once accepted it from the time of their leader, Ptolemy. And yet you, Robert, forgetful of such a consensus, consider me alone and take every opportunity to strike at me from every angle in order to thwart my comparisons.[147]

Kepler described the aspects and consonances as "different peoples, as it were," who came from "the same country of Geometry."[148] Essentially, they originated from the same set of principles in different ways. In the case of the consonances, the section of the circumference of a circle corresponding to the side of an inscribed polygon was extended in a straight line and compared in length with the remaining part of the circumference. Kepler compared this departure from the circle to the foundation of a new colony, where the consonances, "living by their own laws," had established a certain distance from their circular origins.[149] The aspects, on the other hand, were thought to rely completely on the circle for their determination. While the length of a line measured by the side of an inscribed polygon lay at the heart of every consonance, no such feature could be found in the geometrical formulation of an aspect. An aspect

145 Ibid., 258.26–27.

146 Ibid., 234.34.

147 Ibid., 413.10–14: "Diapason et Oppositionis aspectum ex iisdem ego fundamentis Geometricis deduxi: nec huius consociationis author sum primus, omnis astrologorum natio, inde à duce Ptolomaeo dudum eam recepit. At tu Roberte, tanti consensus oblitus, in me solum intueris, me omnibus ictibus per omnes occasiones petis, ut comparationes meas disturbes."

148 Ibid., 261.25–26.

149 Ibid., 261.29–30.

was determined entirely by the inscription of congruent and constructible polygons in a circle. "The aspects, remaining within their own country, the circle," Kepler wrote, "make use of no other laws than those which the roundness of the circle prescribes to them."[150] These figures had been found among the regular plane figures, Kepler wrote, and were "congruent and inscribed in a circle."[151]

Despite their different origins, the aspects and consonances relied similarly on the soul for their recognition. Kepler defined an aspect as "a thing of reason" whose influence could not be conveyed immediately, "as if rain and similar things came down from the heavens themselves," but objectively by an animate faculty.[152] "If there were no soul in the earth," Kepler wrote, the sun, moon, and planets would have no astrological influence, "either on their own or through any suitable aspect."[153] And while the harmonies he identified among the motions of those celestial bodies were not audible, they were thought to resonate with a higher faculty of the soul. In fact, the celestial harmonies involved some of the same relations the soul made instinctually when it enjoyed a musical melody. Kepler claimed that the consonances were not simply created by the fluctuation of the air but consisted more fundamentally in harmonic proportions produced by the human voice as well as the motions of the planets. For the perception of these proportions, Kepler referred to the ability of the soul to identify and appreciate their archetypal essence. In the same way the motions of the planets expressed the archetypal principles of the consonances, Kepler argued, the configuration of the heavens exemplified similar principles that found their resonance in the soul of the earth. "It suffices that there is a soul," Kepler wrote, "which perceives those proportions when they exist and is stirred up by them."[154]

In his critique of the above account, Fludd claimed that the sublunar soul was only part of a more complete system. This was "the very soul of the world," which encompassed every conception of cosmic harmony.[155] Kepler objected that universal participation in harmony did not amount to universal awareness, since only the human soul could understand the geometry that lay at the heart of the influence of the heavens. He claimed the earth could only reach an awareness of those configurations instinctu-

150 Ibid., 261.30–32.
151 Ibid., 261.32–33.
152 Ibid., 240.39–40.
153 Ibid., 241.9–11.
154 Ibid., 450.37–38.
155 Ibid., 451.1.

ally. It reacted to their appearance "like a bull or an elephant," Kepler wrote, which was "slow to ire, yet all the more violently when angered."[156] The human soul, on the other hand, was considered capable of arriving at a precise understanding of the principles that underlay the aspects.

Kepler had expressed sympathy for the view that "the soul of the world [*anima mundi*]" lay in the sun,[157] and he thought Fludd might even agree with him. "You seem to allow for the very same thing," Kepler wrote, "when you refer to the soul of the world as the soul of the sun."[158] Kepler claimed he could accept this idea if Fludd also adopted the Platonic doctrine that situated the human soul "originally in the heart," where it was propelled to the rest of the body by the projection of a *species*.[159] Deployed on a cosmic scale, this doctrine would not allow for the seat of the world soul to lie "either in the air or in the earth," Kepler wrote.[160] To extend "all the way to the fixed stars," it made more sense to situate the soul in the sun, which Kepler thought served as a source of life in the same way as the heart. If others questioned the essential unity of the world soul, Kepler argued that the soul of the earth would still maintain a sense of autonomy while sharing in a greater "unity of essence." To support this view, Kepler turned to human anatomy and compared the essential unity of the organs and their various faculties to "the universal soul of the world." While the earth played a particular part in the cosmic body, Kepler reasoned, it continued to act in concert with "the supremacy of the soul of the world" in fruitful ways:

> Yet shall we rather follow the physicians? However much they may uphold the unity of the soul in man, they identify its diverse faculties according to the diversity of organs, the heart, the liver, and the brain, and even their faculties present in the particular organs. This is clearly also the case in the world, as the sun is a sort of heart and the earth thus plays the part of the liver or spleen. And so there shall be a certain faculty of the earth [that is] innate to it, even if that [faculty] surrenders to the supremacy of the universal soul of the world and submits to intercourse with it on account of [their] unity of essence.[161]

---

[156] Ibid., 268.32–33.
[157] Ibid., 416.25–26.
[158] Ibid., 416.26–27.
[159] Ibid., 416.20–22.
[160] Ibid., 416.23–24.
[161] Ibid., 416.27–34: "An magis medicos sequemur: qui quantumvis unitatem animae in Homine defendant, diversas eius facultates agnoscunt pro diversitate viscerum, Cordis, Epatis, Cerebri, et suas quidem singulis visceribus praesentes facultates? Tunc sane etiam in mundo, ut Sol cor quoddam est, sic Terra Epatis vicem gerit aut lienis. Certa igitur erit Terrae facultas illique insita: quamvis illa principatum cedat Animae mundi universali, eique se per unitatem essentiae copulatam fateatur."

Although the sublunar soul was thought to be only part of the world soul, Kepler claimed it shared in the same fundamental essence.

Fludd also opposed Kepler and his astrometeorology on the basis of the Bible. "In the Sacred Scriptures," Fludd argued, "no evidence is to be found that the weather is excited by the aspects."[162] According to the Book of Genesis, the stars had been created as "lights in the firmament of heaven" to divide the day from the night and serve as signs to measure "the days, the seasons, and the years."[163] Surprised by this selective use of the Bible and opposed to study of it as a source of "scientific authority,"[164] Kepler fired back at Fludd for neglecting natural philosophy. Kepler argued that those who most fiercely rejected the aspects on the absence of biblical evidence were "farthest removed from philosophy."[165] "You may further argue that no mention is made in Scripture of the difference between the wandering and the fixed stars," Kepler wrote wryly, "and so there are no planets."[166] Such a glaring absence, along with those passages that were widely thought to affirm the immobility of the earth, were regarded by Kepler as the result of a natural knowledge far removed from the heights of the seventeenth century. Trained in the humanist techniques of textual criticism, Kepler also believed that reading the Bible was not entirely different from reading any other ancient text.[167] Mindful that it was "a collection of books from ancient history,"[168] Kepler applied his finely tuned philological tools and revolutionary view of astronomy to confirm the compatibility of Sacred Scripture with the new cosmology. When it came to Joshua's command to still the sun while Israel slew the Amorites, for example, Kepler claimed the reader could interpret this passage in another way. "God fulfilled what Joshua wished," he said, "and stopped the motion of the earth in such a way that the sun appeared to stand still."[169] When natural philosophy did not fully square with Sacred Scripture, Kepler argued that biblical language could not always be interpreted physically.[170]

---

162 Ibid., 450.6–7.

163 *Genesis* 1:14–15.

164 On Kepler's opposition to the study of Sacred Scripture as scientific authority, see Howell, 2002, p. 110; cf. Hübner, 1975, pp. 165–175.

165 *JKGW*, 6, 450.7–8.

166 Ibid., 450.11–13.

167 On Kepler's application of humanist analytical techniques to the Bible, see Howell, 2002, and Grafton, 1991, pp. 197–198.

168 Howell, 2002, pp. 116–125.

169 *JKGW*, 3, 30.12–13. Cf. Bieri, 2008, pp. 64–65.

170 On the relation between "the book of the Bible and the book of Nature" in Kepler's natural philosophy, see Hübner, 1975, pp. 158–175.

Differences Over Divinity: Kepler's Final Criticisms of Fludd

In addition to reproaching Fludd for wrongly relying on the Bible, Kepler challenged his choice of ancient authors. "Why should you follow [Hermes] Trismegistus," Kepler asked, "if you forbid me from the company of Plato?"[171] "Why may you employ Iamblichus and Porphyry, enemies of Christian doctrine," he further enquired, "[while] I may not employ Proclus or Aristarchus?"[172] Although Aristarchus and Proclus were not Christian themselves, Kepler accepted their philosophical insight as complementary to a Christian cosmology. At the core of his view, Kepler claimed the world had been ordered geometrically according to "the image of the Trinity."[173] Conceived as a sphere, the cosmos was held to embody the Holy Trinity in "the spherical [surface], the center, and the intervening space."[174] As early as 1595, Kepler had suggested that the sun, situated at the center of the cosmos, represented the Father, while the sphere of the fixed stars represented the Son and the intervening space, made up of infinite equidistant lines extending from the center to the outer sphere, stood for the Holy Spirit.[175] "In the world at rest there are the fixed stars, the sun, and the aura or ethereal intermediary," Kepler wrote, "and in the Trinity there are the Son, the Father, and the [Holy] Spirit."[176] Kepler claimed that the sun lay "in the middle of the planets" and also bore "the image of God the Father" as the source of their motion.[177] "For what creation is for God," he explained, "motion is for the sun."[178]

Kepler based his case for the above claim on geometrical principles he believed were firmly rooted "in the mind of man."[179] He argued that knowledge of these principles was made possible by "our natural aptitude for perceiving geometrical things," which had been given to us as part of

171 Ibid., 451.30–31.

172 Ibid., 451.31–33.

173 Ibid., 453.17–18.

174 *JKGW*, 13, no. 23, 72–73.

175 On the central significance of the sphere in Kepler's Trinitarian conception of the cosmos, see Hübner, 1975, pp. 186–192; Simon, 1979, pp. 133–136; Kozhamthadam, 1994, pp. 16–18; Howell, 2002, pp. 127–129.

176 *JKGW*, 13, no. 23, 73–74.

177 Ibid., 78–80.

178 Ibid., 80. Kepler chose not to discuss Fludd's distribution of the days of creation according to the three persons of the Trinity. "I refer your distribution of the days of creation among the persons of the Holy Trinity to the theologians," he wrote. "It is enough for me if I may show in the very form of the world and its particular parts a certain similarity to the Holy Trinity." See *JKGW*, 6, 441.9–13.

179 Ibid., 456.20.

"the natural participation of the image of God."[180] Kepler applauded Proclus as an early witness of this vision and praised Aristarchus for glimpsing the true structure of the world as well. The archetypal essence of the sun-centered system, he claimed, spoke to the soul in a way that only the human mind could grasp. This was not a matter of choice, Kepler said, since "the architect of the heavens [*architectus coelorum*]" had not followed rules made up by man.[181] God had created the world according to archetypal principles He had taken from His very being.[182] And man, made in His image, found no better basis for knowledge than rediscovering those very principles. "You shall not deny," Kepler wrote to Fludd, "that the geometrical causes are eternal, as reason is itself."[183]

Kepler did not accept the spherical form of the cosmos as a chance discovery or the accidental application of geometrical reasoning. Rather, he saw it as the expression of archetypal principles that were divinely inspired and innate to the very essence of the soul. The principles that appeared in the astrological aspects were thus seen as stemming from "the eternal storehouse of knowledge" along with the very "idea of the heavens [*idea coelorum*]."[184] That knowledge of those principles was possible, Kepler argued, was evident from man's ability to express instinctually and examine rationally the same geometrical archetypes originally employed in the act of creation. It has been observed that Kepler made no real distinction between these principles and the human soul: there was no "soul-and-contents, but only soul," which also became "the harmony itself."[185] Sensible occurrences such as certain configurations of the heavens and the harmonic consonances were thought to awaken us to our own archetypal essence. With this unity in mind, Kepler found at the heart of every form of harmony the same set of divinely inspired principles. "God brought together or gave the same proportions to the motions of the planets that exist in human harmonies," he wrote to Fludd.[186]

Kepler emphasized to Fludd that it was not only the human mind that attested to the archetypes, but their realization in sensible things such as the configurations and the consonances. More than mental principles, the archetypes were manifested materially. To establish a link with the world

---

[180] Ibid., 456.17–21.
[181] Ibid., 456.21–23.
[182] Ibid., 456.23–25.
[183] Ibid., 456.16.
[184] Ibid., 456.14–15.
[185] Escobar, 2008, p. 32.
[186] *JKGW*, 6, 449.28–29.

through the material manifestation of these principles, Kepler suggested that the body conveyed sensory information to the soul, where it could be compared in such a way that produced a sensible harmony.[187]

Although Kepler acknowledged the presence of God in the formal essence of the archetypes, he did not believe their divine nature dictated the meteorological responses of the earth when they appeared in the form of the aspects. Could anyone, he asked, "bind every event of the inferior world by the motions of the stars?"[188] As a living being, the earth was not bound to any behavioral response, nor could Kepler pin any one condition of the weather down to a particular configuration. As he had argued earlier to Fludd, there was neither sympathy nor antipathy between the heavens and the earth, only greater and lesser degrees of influence.[189] Essentially, Kepler conceived of the meteorological responses of the earth as the reverberations of the aspects with the sublunar soul. Resonating rather than ruling, the aspects elicited responses from the earth without determining their result.

To further his case for the sublunar soul, Kepler presented the dismal prospect of a world without it. If no soul existed in the earth, he wrote, there was no reason to accept the birth of anything from it. Described in the Book of Genesis as the source of origin of every land-dwelling animal and plant, the earth was also widely held responsible for the production of meteorological processes. There was a soul in sublunar nature, Kepler argued, "and the word of God the Creator was not in vain when he said, 'Let the earth bring forth.' "[190] Angered by Fludd's assertion that the sublunar soul contradicted Christian doctrine, Kepler claimed that it conflicted in no way with Sacred Scripture. The soul of the earth was also evidenced by what he regarded as a relation between meteorological events and the configurations of the heavens. As Kepler explained, Fludd's criticism of this claim as "a mere fiction, to be accepted in no way by Christians," took no account of the real resonance between certain arrangements of the heavens and the soul of the earth.[191] That this resonance revealed the archetypal rudiments of creation, Kepler claimed, was anything but unchristian.

187 Cf. Escobar, 2008, pp. 32–33.
188 *JKGW*, 6, 451.18–19.
189 Ibid., 413.40–414.2.
190 Ibid., 452.36–38.
191 Ibid., 452.40.

Kepler declared that the influence of the aspects did not deny God's role as "the author of the earth's fruitfulness and produce."[192] If the weather was altered by the aspects, this did not amount to attributing the harvest of the earth to the heavens. The archetypal principles involved in the earthly resonance of the aspects were ultimately found in the formal essence of the Creator. As such, they offered insight into the nature of God and His human image. Kepler identified the divine essence with geometry and argued that it gave the material world an archetypal form. This had provided it with a measure of quantity and structure whose mathematical principles had only recently been revealed by Kepler's revolutionary astronomy. "Without a definite quantity in mind," Kepler wrote, "we discern no unity."[193] Kepler believed he had brought the rediscovery of archetypal principles to an even further frontier in *The Harmony of the World.*

In the end, Kepler condemned Fludd's cosmic harmony as an obscure assembly of "invisible quantities, innumerable numbers," and other obscure wanderings of the English physician.[194] Kepler described "the true astrology" Fludd put forward as "spurious," and he discarded "the soul of world harmony" that gave Fludd such great delight as "imaginary."[195] When it came to Fludd's criticism of *The Harmony of the World* as full of "false chimeras thought up out of thin air," Kepler drew the line at his astronomy.[196] "That may be the case for many things," Kepler wrote, "but when you call the planetary motions 'purely accidental' and on that account pronounce the harmonic proportions drawn from them as 'vain,' that involves an offense to God the Creator."[197] Equally appalling for Kepler was Fludd's failure to establish a common epistemological connection between the human intellect and the natural world. "You say that 'you are enlightened by the superior splendor of knowledge,'" Kepler recalled to Fludd.[198] "To me, that knowledge is unknown."[199]

---

[192] Ibid., 451.15.
[193] Ibid., 431.1–2.
[194] Ibid., 451.37–38.
[195] Ibid., 451.39–41.
[196] Ibid., 443.22–24.
[197] Ibid., 443.24–28.
[198] Ibid., 451.41–452.1.
[199] Ibid., 452.1.

CONCLUSION

Kepler claimed to have arrived at his theory of world harmony on the basis of a new and revolutionary cosmology. In *The New Astronomy*, Kepler had formulated a system of celestial physics he believed could "prove only the opinion of Copernicus right and the two other views [of Ptolemy and Tycho] wrong."[1] This superior form of astronomy had been forged from "the very causes of motion," Kepler claimed, an endeavor that had given him reason to reconsider the nature of planetary motion from the beginning.[2] In his dispute with Fludd, Kepler explained how his causal astronomy had been reinforced by recent advances in observational precision and mathematical demonstration. Focusing on "the visible motions of the planets," Kepler never lost sight of their physical reality as he improved on the powers of perception and prediction.[3] While Fludd may have chanced upon some celestial harmony as the product of pure "invention," Kepler argued, his own astronomy had allowed him to perceive the true music of the heavens.[4] "I may hold the tail by the hand while you embrace the head with your mind," he quipped.[5] Yet Kepler was confident that his grip on the matter was far more firm than what Fludd may have mused mentally.

In this book, I have shown that Kepler continued to consider vitalistic principles in the heavens long after launching his new astronomy. His use of these cosmic principles broadened the arena of natural philosophy on the basis of the heliocentric hypothesis. With the sun considered "for certain" as the central source of motion,[6] the planetary status of the earth suggested the universal application of knowledge acquired closer to home. While this knowledge involved mundane matters, the affairs of the earth served to explain the most spectacular events in the heavens. Clothing the stars in terrestrial attire, Kepler compared the heavens to the earth in order to account for change beyond the sublunar sphere. It was a further step in his synthesis of astronomy and natural philosophy and

---

1 *JKGW*, 3, 20.14–17.
2 Ibid., 20.18–22.
3 *JKGW*, 6, 446.22–26.
4 Ibid., 446.27.
5 Ibid., 446.24–25.
6 *JKGW*, 11,1, 469.39.

the causes at work in the celestial sphere. The causal inventory on which Kepler called involved ideas that originally found a place in a forgotten form of physics. I have focused on some of these ideas with the aim of enriching our understanding of Kepler's cosmology as well as the role of vitalistic principles in early modern science more generally. To this end, I have sought to enhance our view of the period by recapturing the relevance of those principles to a range of issues in astrology and natural philosophy. How did the heavens interact with the earth, and what did their similarity say about the sequence of celestial events at the turn of the seventeenth century? This was a time when "an astonishing succession of novelties in the heavens" shook the very foundations of European society.[7] We cannot arrive at a broader knowledge of their impact without a better understanding of their role in the work of influential authors such as Kepler.

Comets and 'new stars' were deployed by diverse scholars who relished the opportunity to reveal their nature and reconsider (and often reinforce) their own world views.[8] The new star of 1604 raised a series of questions that led Kepler to contemplate the size and structure of the cosmos.[9] What did the new luminary say about the world and the place of the sun in it? Optics was one weapon wielded widely to resolve this question. For a growing number of scholars, it was said to employ the superior accuracy of mathematical demonstration over the "merely probable arguments of the philosophers."[10] While Kepler did not claim a greater degree of certainty for philosophy, he turned to it to suggest the nature and origin of celestial novelty in his causal system. This was one of the "motivating ideas" for his more enduring discoveries in astronomy and we cannot simply sever his physical reasoning as a post-facto failure.[11] Although Kepler was primarily concerned with explaining planetary motion, he could not complete his cosmological synthesis without accounting for celestial mutability. Celestial change carried great weight in what has been called his "full philosophical investigation of the heavens."[12] His explanation of novelty in the heavens shared some of the same "natural faculties" that

[7] Granada, 2009, p. 393.

[8] Bernardino Telesio (1509–1588), for example, referred to the 'new star' of 1572 and the comet of 1577 in his critical response to Aristotle on the nature of comets and the Milky Way. See Telesio, 2012.

[9] Cf. Field, 1988, pp. 19–21.

[10] Jardine, forthcoming.

[11] Krafft, 1991, p. 218. Krafft concentrates "only on Kepler's search for the 'natural' causes" that led to his laws of planetary motion.

[12] Westman, 2011, p. 316.

featured in his account of planetary motion and provoked a mixture of praise and disapproval from his peers.[13] We would not consider these faculties as physical today.

Over the course of his career, Kepler assigned a soul to the earth to explain, among other things, why the weather grew stronger when the heavens were configured a certain way. The sublunar soul was not "a speculative theory" Kepler adapted from ancient authority, but the basis of what he believed was a better meteorology that could be proven empirically.[14] Kepler tied the meteorological activity of the earth to the same system of animate faculties he suggested as a source of comparison to make sense of comets, new stars, and other celestial novelties. Familiarity with these faculties involved areas of physics such as anatomy and biology, which Kepler applied to the ether and every other part of the cosmos. Kepler's account extended beyond planetary motion, a single though significant part of his world picture, to encompass processes of change that completed the causal vision he had first glimpsed in *The New Astronomy*. By accepting this larger vision of his celestial physics, we learn that Kepler did not, as some have claimed, finally "free [it] from any animistic ideas."[15] On the contrary, these ideas remained present, at least potentially, in his mature conception of cosmic harmony.

When it came to astrology,[16] Kepler invited those "learned in natural philosophy" to consider the claims he made "according to physical causes" and communicate their objections.[17] This was an obligation for anyone who made "a serious study of philosophy," Kepler wrote, for it concerned "the worship of God and the welfare of the human race."[18] While improving astrology in this way, Kepler did not aspire to a mechanistic model of celestial influence. Underlying his astrology was a series of archetypal principles he identified with the very essence of God and the original act of creation. These were the "first principles of justification" that unified his different domains of interest and "tied them together in a theology of the created world."[19] Kepler suggested a soul for the recognition of these principles and their resonance on earth. At the same time, he argued that

---

13 *JKGW*, 16, no. 456, 7–10.
14 Schwaetzer, 1997, p. 226.
15 Ibid., p. 296.
16 On the position of astrology as "an integral part" of Kepler's cosmology, see Rabin, 2010, p. 63.
17 *JKGW*, 4, 35.24–27. Cf. Field, 1984b, p. 268.
18 *JKGW*, 4, 35.28–30. Cf. Field, 1984b, p. 268.
19 Westman, 2011, p. 328.

the spontaneous generation of living beings by the sublunar soul shed light on the origins of faraway phenomena in the heavens. Viewed in this way, the creative capacity of the celestial and sublunar spheres accounted for a cycle of decay and renewal according to the same basic principles. This cycle recalled a similar sequence in the sublunar world of Aristotle, where it produced a state of perpetual creation that participated in "the perfection of the universe."[20] The presence of these universal principles in the sensible world allowed the soul to complete an epistemological circle that made knowledge possible and lay at the very core of Kepler's theory of harmony. While we recognize this role of the soul in the music of the heavens, we must also recall the part it played as a powerful analogical resource in the new cosmology.

It has been said that the soul underlay not only Kepler's astrology, "but every [other] dimension of his thought."[21] Whether Kepler viewed it as a vital actor in the earth or analogically as the author of new forms in the ether, the soul was a central part of his cosmology. The vitalistic analogies Kepler applied to the heavens were not empty metaphors, but a fundamental form of knowledge whose explanatory value he took entirely to heart.[22] With this in mind, we do well to remember that our failure to map the faculties of the soul onto the modern world may point us in the direction of appreciating Kepler's vitalistic views more fully. "The coexistence of physical properties and psychic properties in the soul" did not strike Kepler "in the least as a difficulty or problem."[23] When Kepler assigned similar faculties to physical objects in the heavens, their conceptual basis in the living body was never lost entirely. As Kepler explained in *The Epitome of Copernican Astronomy*, the sun acted as a heart that housed a soul and served as "the dispenser of all worldly motion."[24] Just as the human heart spread the soul throughout the body, the soul in the sun—responsible for revolving it—was emitted "together with heat and light" to serve as the central source of motion.[25] Kepler was censured by the Roman Index for committing this "error" of claiming that the sun was

20 Martin, 2011, p. 42.
21 Simon, 1979, p. 195.
22 Ibid., p. 202.
23 Ibid., p. 197.
24 *JKGW*, 7, 354.25–26: "Indeed, I admit a soul in the body of the sun that is responsible for the revolution of the sun and serves as the dispenser of all worldly motion [*totiusque motus mundani dispensatrix*]."
25 Ibid., 354.31–32. Cf. Hübner, 1975, p. 189.

"animated."[26] He had determined that the dense matter of the sun was the ideal dwelling place for a soul that could "master such resilient matter and set it on fire."[27] Together with this fact is the forgotten role of the soul as a source of analogy and metaphor in Kepler's greater enterprise of philosophical inquiry. We cannot afford to cut his cosmology short by severing vitalistic principles from his complete system.

[26] Mayaud, 1997, p. 66. I thank Aviva Rothman for suggesting this source.
[27] *JKGW*, 7, 298.40–41.

BIBLIOGRAPHY

Primary Literature

Original Editions

Fludd, Robert, *Veritatis proscenium, seu Demonstratio quaedam analytica*. Frankfurt: Erasmus Kempffer, 1621.

Hevelius, Johannes, *Cometographia*. Gdansk: Simon Reininger, 1668.

Kepler, Johannes, *De stella nova in pede Serpentarii*. Prague: Paul Sessius, 1606.

——, *Harmonices mundi libri quinque*. Linz: Gottfried Tampach and Johannes Plancus, 1619.

——, *De cometis libelli tres*. Augsburg: Andreas Aperger, 1619–1620.

——, *Mysterium cosmographicum de admirabili proportione orbium coelestium*. Frankfurt: Gottfried Tampach, 1621.

——, *Pro suo opere Harmonices mundi apologia*. Frankfurt: Gottfried Tampach, 1622.

Sacrobosco, Joannis, *Libellus de sphaera: accessit eiusdem autoris computus ecclesiasticus & alia quaedam in studiosorum gratiam edita: cum praefacione Philippi Melanthonis*. Wittenberg: Petrus Seitz, 1543.

Whewell, William, *The Philosophy of the Inductive Sciences, Founded upon Their History*. London: John W. Parker, 1847.

Modern Editions and Translations

Aristotle, *Generation of Animals*. Trans. Arthur L. Peck. Cambridge, MA: Harvard University Press, 1942.

——, *De generatione et corruptione*. Trans. Edward S. Forster and David J. Furley. Cambridge, MA: Harvard University Press, 1965.

——, *De anima*. Trans. Hugh Lawson-Tancred. London: Penguin Books, 1986.

——, *Meteorologica*. Trans. H. D. P. Lee. Cambridge, MA: Harvard University Press, 2004.

——, *On the Heavens*. Trans. W. K. C. Guthrie. Cambridge, MA: Harvard University Press, 2006.

Augustine, *On Free Choice of the Will*. Trans. Thomas Williams. Indianapolis: Hackett, 1993.

Beeckman, Isaac, *Journal tenu par Isaac Beeckman de 1604 à 1634*, vol. 3 (1627–1634). Ed. Cornelis de Waard. The Hague : Martinus Nijhoff, 1945.

Brahe, Tycho, *Opera omnia*. Ed. J. L. E. Dreyer. 15 vols. Copenhagen: Swets & Zeitlinger, 1972.

Cardano, Girolamo, *Opera omnia*, vol. 5. Stuttgart: Friedrich Frommann, 1966.

Cicero, *De divinatione*. Trans. William A. Falconer. London: Hienemann, 1923.

——, *On the Nature of the Gods*. Cambridge, MA: Harvard University Press, 2005.

Comenius, Johann Amos, *Grosse Didaktik*. Trans. Andreas Flitner. Düsseldorf: H. Kupper, 1954.

Hesiod, *Theogony*. Trans. Richmond Lattimore. Ann Arbour: University of Michigan Press, 1994.

Humboldt, Alexander von, *Cosmos: A Sketch of a Physical Description of the Universe*. Trans. E. C. Otté. Baltimore: Johns Hopkins University Press, 1997.

Kepler, Johannes, *Gesammelte Werke*. Ed. Max Caspar and Walther von Dyck *et al.* 22 vols. Munich: C. H. Beck, 1937–.

——, *Conversation with Galileo's* Sidereal Messenger. Trans. Edward Rosen. New York: Johnson, 1965.

——, *The Six-Cornered Snowflake*. Trans. Colin Hardie. Oxford: Clarendon Press, 1966.
——, *The Dream, or Posthumous Work on Lunar Astronomy*. Trans. Edward Rosen. Madison: University of Wisconsin Press, 1967.
——, *The Secret of the Universe*. Trans. A. M. Duncan. New York: Abaris, 1981.
——, *El secreto del universo*. Trans. Eloy Rada García. Madrid: Alianza, 1992.
——, *New Astronomy*. Trans. William H. Donahue. Cambridge: Cambridge University Press, 1992.
——, *The Harmony of the World*. Trans. E. J. Aiton, A. M. Duncan, and J. V. Field. Philadelphia: American Philosophical Society, 1997.
——, *Optics: Paralipomena to Witelo and Optical Part of Astronomy*. Trans. William H. Donahue. Santa Fe: Green Lion Press, 2000.
——, *Vom sechseckigen Schnee*. Trans. Lothar Dunsch. Dresden: Hellerau, 2005.
——, *Über den Neuen Stern im Fuß des Schlangenträgers*. Trans. Otto and Eva Schönberger and Eberhard Knobloch. Würzburg: Königshausen & Neumann, 2006.
——, *Schriften zur Optik, 1604–1611*. Trans. Franz Hammer and Ferdinand Plehn. Frankfurt: Harri Deutsch, 2008.
——, *The Six-Cornered Snowflake: A New Year's Gift*. Trans. Jacques Bromberg. Philadelphia: Paul Dry Books, 2010.
Lucretius, *On the Nature of Things*. London: W. Heinemann, 1924.
Melanchthon, Philipp. *Briefwechsel*, vol. T 5, ed. Walter Thüringer. Stuttgart: Frommann-Holzboog, 2003.
Pico Della Mirandola, Giovanni, *Disputationes adversus astrologiam divinatricem*. Ed./trans. Eugenio Garin. 2 vols. Florence: Vallecchi, 1946–1952.
Plato, *Timaeus and Critias*. Trans. Desmond Lee. London: Penguin Books, 1977.
——, *The Republic*. Trans. Desmond Lee. London: Penguin Books, 1987.
Pliny, *Historia naturalis*. Trans. H. Rackham, D. E. Eichholz, and W. H. S. Jones. 10 vols. London: Heinemann, 1961–1968.
Proclus, *In primum Euclidis Elementorum librum*. Ed. Gottfried Friedlein. Hildesheim: Georg Olms, 1992.
Ptolemy, *Tetrabiblos*. Trans. Frank E. Robbins. London: Heinemann, 1948.
——, *Almagest*. Trans. G. J. Toomer. Princeton: Princeton University Press, 1998.
Seneca, *Naturales quaestiones*. Trans. Thomas H. Corcoran. London: Heinemann, 1972.
——, *Naturales quaestiones*. Trans. Carmen Codoñer. 2 vols. Madrid: Consejo Superior de Investigaciones Científicas, 1979.
Telesio, Bernardino. *Sobre los cometas y la Vía Láctea*. Ed./trans. Miguel Ángel Granada. Madrid: Tecnos, 2012.
Virgil, *Georgics*. Trans. C. Day Lewis. Oxford: Oxford University Press, 1947.

Secondary Literature

Aiton, E. J., *The Vortex Theory of Planetary Motions*. New York: American Elsevier, 1972.
——, "Johannes Kepler and the Astronomy without Hypotheses." In: *Japanese Studies in the History of Science* 14 (1975), 49–72.
Armitage, Angus. *John Kepler*. London: Faber and Faber, 1966.
Barker, Peter, "Stoic Contributions to Early Modern Science." In: *Atoms,* Pnuema, *and Tranquillity: Epicurean and Stoic Themes in European Thought*, ed. Margaret J. Osler (Cambridge: Cambridge University Press, 1991), 135–154.
——, "The Optical Theory of Comets from Apian to Kepler." In: *Physis* 30 (1993), 1–25.
——, "Kepler's Epistemology." In: *Method and Order in Renaissance Philosophy*, ed. D. A. Di Liscia, E. Kessler, and C. Methuen (Aldershot: Ashgate, 1997), 355–368.
——, "Copernicus and the Critics of Ptolemy." In: *Journal for the History of Astronomy* 30 (1999), 343–357.
——, "The Role of Religion in the Lutheran Response to Copernicus." In: *Rethinking the Scientific Revolution*, ed. Margaret J. Osler (Cambridge: Cambridge University Press, 2000), 59–88.

——, "Theological Foundations of Kepler's Astronomy." In: *Osiris* 16 (2001), 88–113.
——, "Astronomy, Providence, and the Lutheran Contribution to Science." In: *Reading God's World: The Scientific Vocation*, ed. Angus J. L. Menuge (St Louis: Concordia Publishing House, 2004), 157–187.
——, "New Work in Early Modern Science." In: *Centaurus* 48 (2006), 1–2.
Barker, Peter, and Bernard R. Goldstein, "Realism and Instrumentalism in Sixteenth Century Astronomy: A Reappraisal." In: *Perspectives on Science* 6 (1998), 232–258.
Baumgardt, Carola, *Johannes Kepler: Life and Letters*. New York: Philosophical Library, 1951.
Beer, Arthur, "Kepler's Astrology and Mysticism." In: *Vistas in Astronomy* 18 (1975), 399–426.
Berkel, Klaas Van, *Isaac Beeckman (1588–1637) en de mechanisering van het wereldbeeld.* Amsterdam: Rodopi, 1983.
Bialas, Volker, *Johannes Kepler*. Munich: C. H. Beck, 2004.
Bieri, Hans. *Der Streit um das kopernikanische Weltsystem im 17. Jahrhundert: Galileo Galileis Akkomodationstheorie und ihre historischen Hintergründe.* Bern: Peter Lang, 2008.
Boner, Patrick J., "Soul-Searching with Kepler: An Analysis of *Anima* in His Astrology." In: *Journal for the History of Astronomy* 36 (2005), 7–20.
——, "Kepler's Living Cosmology: Bridging the Celestial and Terrestrial Realms." In: *Centaurus* 48 (2006a), 32–39.
——, "Kepler on the Origins of Comets: Applying Earthly Knowledge to Celestial Events." In: *Nuncius* 21 (2006b), 31–47.
——, "Kepler v. the Epicureans: Causality, Coincidence and the Origins of the New Star of 1604." In: *Journal for the History of Astronomy* 38 (2007), 207–221.
——, "Kepler's Early Astrological Calendars: Matter, Methodology and Multidisciplinarity." *Centaurus* 50 (2008a), 324–328.
——, "Life in the Liquid Fields: Kepler, Tycho and Gilbert on the Nature of the Heavens and Earth." In: *History of Science* 46 (2008b), 275–297.
——, "Finding Favour in the Heavens and Earth: Stadius, Kepler and Astrological Calendars in Early Modern Graz." In: *Studia Copernicana* 42 (2009a), 159–177.
——, "A Tenuous Tandem: Patrizi and Kepler on the Origins of Stars." In: *Journal for the History of Astronomy* 40 (2009b), 381–391.
——, "Astrology on Trial: Kepler, Pico and the Preservation of the Aspects." In: *Culture and Cosmos* 14 (2010), 209–234.
——, "Kepler's Copernican Campaign and the New Star of 1604." In: *Change and Continuity in Early Modern Cosmology*, ed. Patrick J. Boner (Dordrecht: Springer, 2011a), 93–114.
——, "Kepler's Imprecise Astrology." In: *Kepler: La physique céleste*, ed. Édouard Mehl (Paris: Les Belles Lettres, 2011b), 115–132.
——, "Kepler's Vitalistic View of the Heavens: Some Preliminary Remarks." In: *Novas y cometas entre 1572 y 1618: Revolución cosmológica y renovación política y religiosa*, ed. Miguel Ángel Granada (Barcelona: University of Barcelona, 2012a), 165–193.
——, "Beached Whales and Priests of God: Kepler and the Cometary Spirit of 1607." In: *Early Science and Medicine* 17 (2012b), 589–603.
Boockmann, Friederike, "Die Horoskopsammlung von Johannes Kepler." In: *Miscellanea Kepleriana: Festchrift für Volker Bialas zum 65. Geburtstag*, ed. Friederike Boockmann, Daniel A. Di Liscia, and Hella Kothmann (Augsburg: Dr. Erwin Rauner, 2005), 183–203.
Bouché-Leclercq, Auguste, *L'astrologie grecque*. Paris: E. Leroux, 1899.
Bowden, Mary Ellen, *The Scientific Revolution in Astrology: The English Reformers, 1558–1686*. Ph.D. Dissertation. Yale University, 1984.
Brackenridge, J. Bruce, "Kepler, Elliptical Orbits and Celestial Circularity: A Study in the Persistence of Metaphysical Commitment (1)." In: *Annals of Science* 39 (1982), 117–143.
Brackenridge, J. Bruce, and Mary Ann Rossi, "Johannes Kepler's *On the More Certain Fundamentals of Astrology*, Prague 1601." In: *Proceedings of the American Philosophical Society* 123 (1979), 85–116.

Burke-Gaffney, Michael W., *Kepler and the Jesuits*. Milwaukee: Bruce Publishing Company, 1944.

Bucciantini, Massimo, *Galileo e Keplero: Filosofia, cosmologia e teologia nell'Età della Controriforma*. Turin: Einaudi, 2003.

Burtt, Edwin A., *The Metaphysical Foundations of Modern Physical Science*. Atlantic Highlands: Humanities Press, 1932.

Butterfield, Herbert, *The Origins of Modern Science 1300–1800*. New York: Free Press, 1957.

Campion, Nicholas, *A History of Western Astrology, II: The Medieval and Modern Worlds*. London, Continuum, 2009.

Caroti, Stefano, "Melanchthon's Astrology." In: "Astrologi hallucinati": *Stars and the End of the World in Luther's Time*, ed. Paola Zambelli (Berlin: Walter de Gruyter, 1986), 109–121.

Caspar, Max, *Kepler*. Ed./trans. C. Doris Hellman. New York: Dover, 1993.

Christianson, John R., "Tycho Brahe's German Treatise on the Comet of 1577: A Study in Science and Politics." In: *Isis* 70 (1979), 110–140.

——, *On Tycho's Island: Tycho Brahe and His Assistants, 1570–1601*. Cambridge: Cambridge University Press, 2000.

Cohen, H. Floris., *The Scientific Revolution: A Historiographical Inquiry*. Chicago: University of Chicago Press, 1994.

Coopland, G. W., *Nicole Oresme and the Astrologers: A Study of His* Livre de divinacions. Cambridge, MA: Harvard University Press, 1952.

Curry, Patrick, *Prophecy and Power: Astrology in Early Modern England*. Princeton: Princeton University Press, 1989.

——, "The Historiography of Astrology: A Diagnosis and a Prescription." In: *Horoscopes and Public Spheres: Essays on the History of Astrology*, ed. Günther Oestmann, H. Darrel Rutkin, and Kocku von Stuckrad (Berlin: Walter de Gruyter, 2005), 261–274.

Davis, A. E. L., "Kepler's *via ovalis composita*: Unity from Diversity." In: *Journal for the History of Astronomy* 40 (2009), 55–69.

Dijksterhuis, Eduard J., *The Mechanization of the World Picture: Pythagoras to Newton*. Trans. C. Dikshoorn. Oxford: Clarendon Press, 1961.

Di Liscia, Daniel A., "Pneumatica." In: *Johannes Kepler Gesammelte Werke*, vol. 21,2.2, ed. Friederike Boockmann and Daniel A. Di Liscia (Munich: C. H. Beck, 2009), 667–691.

Dixon, C. Scott, "Popular Astrology and Lutheran Propaganda in Reformation Germany." In: *History* 84 (1999), 403–418.

Donahue, William H., "Philosophy in Kepler's Astronomy." In: *Journal for the History of Astronomy* 33 (2002), 296–297.

Drake, Stillman, and Charles D. O'Malley, *The Controversy on the Comets of 1618: Galileo Galilei, Horatio Grassi, Mario Guiducci, Johann Kepler*. Philadelphia: University of Pennsylvania Press, 1960.

Dreyer, J. L. E., *A History of Astronomy from Thales to Kepler*. New York: Dover, 1953.

Eichholz, D. E., "Aristotle's Theory of the Formation of Metals and Minerals." In: *The Classical Quarterly* 43 (1949), 141–146.

Escobar, Jorge Manuel, "Kepler's Theory of the Soul: A Study on Epistemology." In: *Studies in History and Philosophy of Science* 39 (2008), 15–41.

Evans, R. J. W., *Rudolf II and His World: A Study in Intellectual History, 1576–1612*. Oxford: Oxford University Press, 1973.

Fabbri, Natacha, *Cosmologia e armonia in Kepler e Mersenne: Contrappunto a due voci sul tema dell'*Harmonice mundi. Florence: Leo S. Olschki, 2003.

——, *De l'utilité de l'harmonie: Filosofia, scienza e musica in Mersenne, Descartes e Galileo*. Pisa: Scuola Normale Superiore, 2008.

Field, Judith V., "Kepler's Rejection of Numerology." In: *Occult and Scientific Mentalities in the Renaissance*, ed. Brian Vickers (Cambridge: Cambridge University Press, 1984a), 272–296.

——, "A Lutheran Astrologer: Johannes Kepler." In: *Archive for History of Exact Sciences* 31 (1984b), 189–272.

——, *Kepler's Geometrical Cosmology*. London: Athlone Press, 1988.

Fleckenstein, Joachim Otto, "Kepler and Neoplatonism." In: *Vistas in Astronomy* 18 (1975), 427–435.

Garin, Eugenio, *Astrology in the Renaissance: The Zodiac of Life*. Trans. Clare Robertson. London: Routledge and Kegan Paul, 1983.

Geneva, Ann, *Astrology and the Seventeenth Century Mind: William Lilly and the Language of the Stars*. New York: St. Martin's Press, 1995.

Giglioni, Guido, "Panpsychism versus Hylozoism: An Interpretation of Some Seventeenth-Century Doctrines of Universal Animation." In: *Acta Comeniana* 11 (1995), 25–45.

Gingerich, Owen, "Johannes Kepler and the *Rudolphine Tables*." In: *Sky and Telescope* 21 (1971), 328–333.

——, "Kepler's Place in Astronomy." *Vistas in Astronomy* 18 (1975), 261–278.

——, "Johannes Kepler." In: *Dictionary of Scientific Biography*, ed. Charles C. Gillispie (New York: Scribner, 1981), 289–310.

——, *The Eye of Heaven: Ptolemy, Copernicus, Kepler*. New York: American Institute of Physics, 1993.

——, "Kepler Then and Now." In: *Perspectives on Science* 10 (2002), 228–240.

——, *The Book Nobody Read: Chasing the Revolutions of Nicolaus Copernicus*. New York: Walker, 2004.

——, "The Delights of a Roving Mind." In: *The Six-Cornered Snowflake: A New Year's Gift*, Johannes Kepler, trans. Jacques Bromberg (Philadelphia: Paul Dry Books, 2010), 3–11.

Goddu, André, *Copernicus and the Aristotelian Tradition: Education, Reading, and Philosophy in Copernicus's Path to Heliocentrism*. Leiden: Brill, 2010.

Grafton, Anthony, *Defenders of the Text: The Traditions of Scholarship in an Age of Science, 1450–1800*. Cambridge, MA: Harvard University Press, 1991.

——, "Kepler as a Reader." In: *Journal of the History of Ideas* 53 (1992), 561–572.

——, *Commerce with the Classics: Ancient Books and Renaissance Readers*. Ann Arbor: University of Michigan Press, 1997.

——, *Cardano's Cosmos: The Worlds and Works of a Renaissance Astrologer*. Cambridge, MA: Harvard University Press, 2000.

Granada, Miguel Ángel, *El debate cosmológico en 1588: Bruno, Brahe, Rothmann, Ursus, Röslin*. Naples: Istituto Italiano per gli Studi Filosofici, 1996.

——, "Cálculos cronológicos, novedades cosmológicas y expectativas escatológicas en la Europa del siglo XVI." In: *Rinascimento*, 2nd ser., 37 (1997), 357–435.

——, *Sfere solide e cielo fluido: Momenti del dibattito cosmologico nella seconda metà del cinquecento*. Naples: Guerini e Associati, 2002.

——, "Kepler v. Roeslin on the Interpretation of Kepler's Nova: (1) 1604–1606." In: *Journal for the History of Astronomy* 36 (2005), 299–319.

——, "Did Tycho Eliminate the Celestial Spheres before 1586?" In: *Journal for the History of Astronomy* 37 (2006a), 125–145.

——, "Helisaeus Röslin on the Eve of the Appearance of the Nova of 1604: His Eschatological Expectations and His Intellectual Career as Recorded in the '*Ratio studiorum et operum meorum*' (1603–1604)." In: *Sudhoffs Archiv* 90 (2006b), 75–96.

——, "Michael Maestlin and the New Star of 1572." In: *Journal for the History of Astronomy* 38 (2007), 99–124.

——, "Kepler and Bruno on the Infinity of the Universe and of Solar Systems." In: *Journal for the History of Astronomy* 39 (2008), 469–495.

——, "Novelties in the Heavens between 1572 and 1604 and Kepler's Unified View of Nature." In: *Journal for the History of Astronomy* 40 (2009), 393–402.

——, " 'A quo moventur planetae?' Kepler et la question de l'agent du movement planétaire après la disparition des orbes solides." In: *Galilaeana* 7 (2010), 111–141.

——, "Johannes Kepler and David Fabricius: Their Discussion on the Nova of 1604." In: *Change and Continuity in Early Modern Cosmology*, ed. Patrick J. Boner (Dordrecht: Springer, 2011a), 67–92.

——, "Roeslin and Kepler: (2) Their Discussion on the Significance of the Celestial Novelties (1607–1613)." In: *Journal for the History of Astronomy* 42 (2011b), 353–390.

Grant, Edward (ed.), *Nicole Oresme and the Kinematics of Circular Motion:* Tractatus de commensurabilitate vel incommensurabilitate motuum celi. Madison: University of Wisconsin Press, 1971.

——, "Celestial Orbs in the Middle Ages." In: *Isis* 78 (1987), 153–173.

Haase, Rudolf, "Kepler's Harmonies, between *pansophia* and *mathesis universalis*." In: *Vistas in Astronomy* 18 (1975), 519–533.

Hallyn, Fernand, *The Poetic Structure of the World: Copernicus and Kepler*. Trans. Donald M. Leslie. New York: Zone Books, 1990.

Hatch, Robert A., "Discovering Mira Ceti: Celestial Change and Cosmic Continuity." In: *Change and Continuity in Early Modern Cosmology*, ed. Patrick J. Boner (Dordrecht: Springer, 2011), 153–176.

Hellman, C. Doris, *The Comet of 1577: Its Place in the History of Astronomy*. New York: AMS Press, 1971.

Henry, John, *The Scientific Revolution and the Origins of Modern Science*. Basingstoke and New York: Palgrave Macmillan, 2008.

Hirai, Hiro, "Concepts of Seeds and Nature in the Work of Marsilio Ficino." In: *Marsilio Ficino: His Theology, His Philosophy, His Legacy*, ed. Michael J. B. Allen and Valery Rees (Leiden: Brill, 2002), 257–284.

——, *Medical Humanism and Natural Philosophy: Renaissance Debates on Matter, Life and the Soul*. Leiden: Brill, 2011.

Holton, Gerald, "Johannes Kepler's Universe: Its Physics and Metaphysics." *American Journal of Physics* 24 (1956), 340–351.

Hooykaas, Reijer, *Religion and the Rise of Modern Science*. Edinburgh: Scottish Academic Press, 1972.

——, "The Rise of Modern Science: When and Why?" In: *The British Journal for the History of Science* 20 (1987), 453–473.

Hotson, Howard, *Johann Heinrich Alsted (1588–1638): Between Renaissance, Reformation, and Universal Reform*. Oxford: Oxford University Press, 2000.

Howell, Kenneth J., *God's Two Books: Copernican Cosmology and Biblical Interpretation in Early Modern Science*. Notre Dame: University of Notre Dame Press, 2002.

Hübner, Jürgen, *Die Theologie Johannes Keplers zwischen Orthodoxie und Naturwissenschaft*. Tübingen: Mohr, 1975.

Huffman, William H., *Robert Fludd and the End of the Renaissance*. London: Routledge, 1988.

Illmer, Markus M., *Die göttliche Mathematik Johannes Keplers: Zur ontologischen Grundlegung des naturwissenschaftlichen Weltbildes*. St. Ottilien: EOS, 1991.

Itokazu, Anastasia Guidi, "A força que move os planetas: Da noção de 'species immateriata' na astronomia de Johannes Kepler." In: *Cadernos de história e filosofía da ciencia* 16 (2006), 211–231.

——, "Da potência motriz solar kepleriana como emanação imaterial." In: *Cadernos de história e filosofía da ciencia* 17 (2007), 303–324.

Jardine, Nicholas, *The Birth of History and Philosophy of Science: Kepler's* A Defence of Tycho against Ursus *with Essays on Its Provenance and Significance*. Cambridge: Cambridge University Press, 1984.

——, "The Places of Astronomy in Early-Modern Culture." In: *Journal for the History of Astronomy* 29 (1998), 49–62.

——, "Koyré's Kepler / Kepler's Koyré." In: *History of Science* 38 (2000), 363–376.

——, "Kepler, God, and the Virtues of Copernican Hypotheses." In: *Nouvaeu ciel, nouvelle terre: La révolution copernicienne dans l'Allemagne de la Réforme (1530–1630)*, ed. Miguel A. Granada and Édouard Mehl (Paris: Les Belles Lettres, 2009), 269–281.

——, "How to Present a Copernican Comet: The Form and Tactics of Christoph Rothmann's *Dialexis* on the Comet of 1585." In: *Celestial Novelties on the Eve of the Scientific Revolution*, ed. Patrick J. Boner and Dario Tessicini (Florence: Olschki, forthcoming).

Jardine, Nicholas, and Alain-Philippe Segonds, "Kepler as Reader and Translator of Aristotle." In: *Philosophy in the Sixteenth and Seventeenth Centuries: Conversations with Aristotle*, ed. Constance Blackwell and Sachiko Kusukawa (Aldershot: Ashgate, 1999), 206–233.

——, *La guerre des astronomes: La querelle au sujet de l'origine du système géo-héliocentrique à la fin du XVI*[e] *siècle*. 2 vols. Paris: Les Belles Lettres, 2008.

Jervis, Jane L., *Cometary Theory in Fifteenth-Century Europe*. Dordrecht: D. Riedel, 1985.

Keller, David R., and E. Charles Brummer, "Putting Food Production in Context: Toward a Postmechanistic Agricultural Ethic." In: *Bioscience* 52 (2002), 264–271.

Kjellgren, Martin, *Taming the Prophets: Astrology, Orthodoxy and the Word of God in Early Modern Sweden*. Lund: Sekel, 2011.

Koestler, Arthur, *The Watershed: A Biography of Johannes Kepler*. New York: Anchor Books, 1960.

Kothmann, Hella, "Der halsstarrige Mars sucht sich eine Wohnung: Johannes Kepler an Kurfürst Christian II. von Sachsen—eine Wiederentdeckung." In: *Beiträge zur Astronomiegeschichte* 11 (2011), 164–172.

Koyré, Alexandre, *From the Closed World to the Infinite Universe*. Baltimore: Johns Hopkins Press, 1957.

——, *Newtonian Studies*. London: Chapman & Hall, 1965.

——, *The Astronomical Revolution: Copernicus—Kepler—Borelli*. Trans. R. E. W. Maddison. New York: Dover, 1992.

Kozhamthadam, Job, *The Discovery of Kepler's Laws: The Interaction of Science, Philosophy, and Religion*. Notre Dame: University of Notre Dame Press, 1994.

Krafft, Fritz, "The New Celestial Physics of Johannes Kepler." In: *Physics, Cosmology and Astronomy, 1300–1700: Tension and Accommodation*, ed. Sabetai Unguru (Dordrecht: Kluwer, 1991), 185–227.

Kuhn, Thomas S., *The Copernican Revolution: Planetary Astronomy in the Development of Western Thought*. Cambridge, MA: Harvard University Press, 1957.

——, *The Structure of Scientific Revolutions*. Chicago: University of Chicago Press, 1996.

Kusukawa, Sachiko, *The Transformation of Natural Philosophy: The Case of Philip Melanchthon*. Cambridge: Cambridge University Press, 1995.

Launert, Dieter, "Le système du monde de Nicolas Reymer Ursus comparé à ceux de Tycho Brahé et Helisaeus Röslin." In: *Kepler: La physique céleste*, ed. Édouard Mehl (Paris: Les Belles Lettres, 2011), 97–111.

Launert, Dieter, and Wolfgang Müller, *Nicolaus Reymers: Ein berühmter Mathematiker und Naturwissenschaftler aus Dithmarschen und seine* Arithmetica analytica. Meldorf: Meldorfer Gelehrtenschule, 1994.

Lerner, Michel-Pierre, "Le problème de la matière céleste après 1550: Aspects de la bataille des cieux fluids." In: *Revue d'histoire des sciences* 42 (1989), 255–280.

——, "La supernova de 1572: Une diversité d'interprétations." In: *L'astronomie* 119 (2005), 558–565.

Lindberg, David C., *Theories of Vision from Al-Kindi to Kepler*. Chicago: University of Chicago Press, 1976.

——, "The Genesis of Kepler's Theory of Light: Light Metaphysics from Plotinus to Kepler." In: *Osiris*, 2nd ser., 2 (1986), 5–42.

——, "Kepler and the Incorporeality of Light." In: *Physics, Cosmology and Astronomy, 1300–1700: Tension and Accommodation*, ed. Sabetai Unguru (Dordrecht: Kluwer, 1991), 229–250.

Luminet, Jean-Pierre, *El tesoro de Kepler*. Barcelona: Ediciones B, 2009.

Magruder, Kerry V., "Global Visions and the Establishment of Theories of the Earth." In: *Centaurus* 48 (2006), 234–257.

Mahnke, Dietrich, *Unendliche Sphäre und Allmittelpunkt*. Halle: Niemeyer, 1937.

Maier, Anneliese, *Die Mechanisierung des Weltbildes im 17. Jahrhundert*. Leipzig: Meiner, 1938.

Martens, Rhonda S., *Kepler's Philosophy and the New Astronomy*. Princeton: Princeton University Press, 2000.

——, "A Commentary on Genesis: Plato's *Timaeus* and Kepler's Astronomy." In: *Plato's* Timaeus *as Cultural Icon*, ed. Gretchen J. Reydams-Schils (Notre Dame: University of Notre Dame Press, 2003), 251–266.

Martin, Craig, *Renaissance Meteorology: Pomponazzi to Descartes*. Baltimore: Johns Hopkins University Press, 2011.

Maurice, Klaus, "Jost Bürgi, or On Innovation." In: *The Clockwork Universe: German Clocks and Automata, 1550–1650*, ed. Klaus Maurice and Otto Mayr (New York: Neale Watson, 1980), 87–102.

Mayaud, Pierre-Noël, *La condamnation des livres coperniciens et sa révocation*. Rome: Editrice Pontificia Università Gregoriana, 1997.

Mayr, Otto, "A Mechanical Symbol for an Authoritarian World." In: *The Clockwork Universe: German Clocks and Automata, 1550–1650*, ed. Klaus Maurice and Otto Mayr (New York: Neale Watson, 1980), 1–8.

——, *Authority, Liberty, and Automatic Machinery in Early Modern Europe*. Baltimore: Johns Hopkins University Press, 1989.

Merchant, Carolyn, *The Death of Nature: Women, Ecology, and the Scientific Revolution*. New York: Harper Collins, 1989.

Methuen, Charlotte, *Kepler's Tübingen: Stimulus to a Theological Mathematics*. Aldershot: Ashgate, 1998.

——, "Special Providence and Sixteenth-Century Astronomical Observation." In: *Early Science and Medicine* 4 (1999), 99–113.

Mosley, Adam J., *Bearing the Heavens: Tycho Brahe and the Astronomical Community of the Late Sixteenth Century*. Cambridge: Cambridge University Press, 2007.

——, "Heaven and Earth in the Late-Sixteenth Century: Tycho and Kepler on the Sub- and Supra-lunary." In: *Nouveau ciel, nouvelle terre: La révolution copernicienne dans l'Allemagne de la Réforme (1530–1630)*, ed Miguel A. Granada and Édouard Mehl (Paris: Les Belles Lettres, 2009), 143–154.

Muir-Harmony, Teasel, David H. DeVorkin, and Peter Abrahams, "Wide-Field Photographic Telescopes: The Yale, Harvard and Harvard/Smithsonian Meteor and Satellite Camera Networks." In: *From Earth-Bound to Satellite: Telescopes, Skills and Networks*, ed. Alison D. Morrison-Low, Sven Dupré, Stephen Johnston, and Giorgio Strano (Leiden: Brill, 2012), 195–234.

Neumann, Hanns-Peter, "*Machina Machinarum*: Die Uhr als Begriff und Metapher zwischen 1450 und 1750." In: *Early Science and Medicine* 15 (2010), 122–191.

Newman, William R., *Atoms and Alchemy: Chymistry and the Experimental Origins of the Scientific Revolution*. Chicago: University of Chicago Press, 2006.

North, John, *God's Clockmaker: Richard of Wallingford and the Invention of Time*. London: Continuum, 2005.

Oelschlaeger, Max, *The Idea of Wilderness: From Prehistory to the Age of Ecology*. New Haven: Yale University Press, 1991.

Omodeo, Pietro Daniel, "David Origanus's Planetary System (1599 and 1609)." In: *Journal for the History of Astronomy* 42 (2011), 439–454.

Osler, Margaret J., *Reconfiguring the World: Nature, God, and Human Understanding from the Middle Ages to Early Modern Europe*. Baltimore: Johns Hopkins University Press, 2010.

Pauli, Wolfgang, "The Influence of Archetypal Ideas on the Scientific Theories of Kepler." In: *The Interpretation of Nature and the Psyche*, ed. Carl G. Jung and Wolfgang Pauli (London: Routledge and Kegan Paul, 1952), 151–247.

Principe, Lawrence M., *The Scientific Revolution: A Very Short Introduction*. Oxford: Oxford University Press, 2011.

Rabin, Sheila J., *Two Renaissance Views of Astrology: Pico and Kepler*. Ph.D. Dissertation. City University of New York, 1987.
——, "Kepler's Attitude toward Pico and the Anti-Astrology Polemic." In: *Renaissance Quarterly* 50 (1997), 750–770.
——, "Was Kepler's *species immateriata* Substantial?" In: *Journal for the History of Astronomy* 36 (2005), 49–56.
——, "The Astrological Cosmos of Johannes Kepler." In: *World-Building and the Early Modern Imagination*, ed. Allison B. Kavey (New York: Palgrave Macmillan, 2010), 59–66.
Rosen, Edward, "Kepler and the Lutheran Attitude towards Copernicanism in the Context of the Struggle between Science and Religion." In: *Vistas in Astronomy* 18 (1975), 399–426.
——, "Kepler's Attitude toward Astrology and Mysticism." In: *Occult and Scientific Mentalities in the Renaissance*, ed. Brian Vickers (Cambridge: Cambridge University Press, 1984), 253–271.
——, "The Dissolution of the Celestial Spheres." In: *Journal for the History of Ideas* 46 (1985), 13–31.
——, *Three Imperial Mathematicians: Kepler Trapped between Tycho Brahe and Ursus*. New York: Abaris, 1986.
Rossi, Paolo, *La scienza e la filosofia dei moderni: Aspetti della rivoluzione scientifica*. Turin: Bollati Boringhieri, 1989.
Rothman, Aviva, "Forms of Persuasion: Kepler, Galileo, and the Dissemination of Copernicanism." In: *Journal for the History of Astronomy* 40 (2009), 403–419.
——, "From Cosmos to Confession: Kepler and the Connection between Astronomical and Religious Truth." In: *Change and Continuity in Early Modern Cosmology*, ed. Patrick J. Boner (Dordrecht: Springer, 2011), 115–133.
Ruffner, James. A., "The Curved and the Straight: Cometary Theory from Kepler to Hevelius." In: *Journal for the History of Astronomy* 2 (1971), 178–194.
Samsonow, Elisabeth von, *Die Erzeugung des Sichtbaren: Die philosophische Begründung naturwissenschaftlicher Wahrheit bei Johannes Kepler*. Munich: Fink, 1986.
Schneer, Cecil, "Kepler's New Year's Gift of a Snowflake." In: *Isis* 51 (1960), 531–545.
Schuster, Robert L., and Lynn M. Highland, "Urban Landslides: Socioeconomic Impacts and Overview of Mitigative Strategies." In: *Bulletin of Engineering Geology and the Environment* 66 (2007), 1–27.
Schwaetzer, Harald,"Si nulla esset in terra anima": *Johannes Keplers Seelenlehre als Grundlage seines Wissenschaftsverständnisses*. Zurich: Georg Olms, 1997.
Segonds, Alain-Philippe, "Tycho Brahe et l'alchimie." In: *Alchimie et philosophie à la Renaissance*, ed. Jean-Claude Margolin and Sylvain Matton (Paris: J. Vrin, 1993), 365–378.
Shackelford, Jole R., "Providence, Power, and Cosmic Causality in Early Modern Astronomy: The Case of Tycho Brahe and Petrus Severinus." In: *Tycho Brahe and Prague: Crossroads of European Science*, ed. John R. Christianson, Alena Hadravová, Petr Hadrava, and Martin Šolc (Frankfurt: H. Deutsch, 2002), 46–69.
Shapin, Steven, *The Scientific Revolution*. Chicago: University of Chicago Press, 1996.
Simon, Gérard, "Kepler's Astrology: The Direction of a Reform." In: *Vistas in Astronomy* 18 (1975), 439–448.
——, *Kepler astronome astrologue*. Paris: Gallimard, 1979.
Smith, A. Mark, "Knowing Things Inside Out: The Scientific Revolution from a Medieval Perspective." In: *The American Historical Review* 95 (1990), 726–744.
Solís, Carlos, "Los cometas contra Copérnico: Brahe, Galileo y los Jesuitas." In: *Galileo y la gestación de la ciencia moderna*, ed. G. Lopez and A. Alejandre (Las Palmas: Fundación Canaria Orotava de Historia de la Ciencia, 2001), 49–62.
Steneck, Nicholas H., *Science and Creation in the Middle Ages: Henry of Langenstein (d. 1397) on* Genesis. Notre Dame: University of Notre Dame Press, 1976.
Stephenson, Bruce, *Kepler's Physical Astronomy*. New York: Springer, 1987.

——, *The Music of the Heavens: Kepler's Harmonic Astronomy*. Princeton: Princeton University Press, 1994.
Straker, Stephen, *Kepler's Optics: A Study in the Development of Seventeenth-Century Natural Philosophy*. Ph.D. Dissertation. Indiana University, 1971.
Sutter, Alex, *Göttliche Maschinen: Die Automaten für Lebendiges bei Descartes, Leibniz, La Mettrie und Kant*. Frankfurt: Athenäum, 1989.
Sutter, Berthold, "Johannes Keplers Stellung innerhalb der Grazer Kalendartradition des 16. Jahrhunderts." In: *Johannes Kepler 1571–1971: Gedenkschrift der Universität Graz*, ed. Paul Urban and Berthold Sutter (Graz: Leykam, 1975), 209–373.
Taub, Liba C., *Ptolemy's Universe: The Natural Philosophical and Ethical Foundations of Ptolemy's Astronomy*. Chicago: Open Court, 1993.
——, *Ancient Meteorology*. London: Routledge, 2003.
Tessicini, Dario, *I dintorni dell'infinito: Giordano Bruno e l'astronomia del Cinquecento*. Pisa and Rome: Fabrizio Serra, 2007.
Tester, S. Jim, *A History of Western Astrology*. Bury St Edmunds: Boydell Press, 1987.
Thorndike, Lynn, *A History of Magic and Experimental Science*, vol. 5. New York: Columbia University Press, 1934.
Tondorf, E. A., "Kepler's Attitude toward Astrology." *Popular Astronomy* 12 (1904), 303.
Vanden Broecke, Steven, *The Limits of Influence: Pico, Louvain, and the Crisis of Renaissance Astrology*. Leiden: Brill, 2003.
Vermij, Rienk, *The Calvinist Copernicans: The Reception of the New Astronomy in the Dutch Republic, 1575–1750*. Amsterdam: Koninklijke Nederlandse Akademie van Wetenschappen, 2002.
——, "Putting the Earth in Heaven: Philips Lansbergen, the Early Dutch Copernicans and the Mechanization of the World Picture." In: *Mechanics and Cosmology in the Medieval and Early Modern Period*, ed. Massimo Bucciantini, Michele Camerota, and Sophie Roux (Florence: Leo S. Olschki, 2007), 121–141.
Vernet, Juan, "Kepler y los horóscopos de Mahoma y Lucero." In: *Al-Andalus: Revista de las escuelas de estudios árabes de Madrid y Granada* 37 (1972), 453–462.
Voelkel, James R., *Johannes Kepler and the New Astronomy*. Oxford: Oxford University Press, 1999a.
——, "Publish or Perish: Legal Contingencies and the Publication of Kepler's *Astronomia nova*." In: *Science in Context* 12 (1999b), 35–59.
——, *The Composition of Kepler's* Astronomia nova. Princeton: Princeton University Press, 2001.
Westman, Robert S., *Johannes Kepler's Adoption of the Copernican Hypothesis*. Ph.D. Dissertation. University of Michigan, 1971.
——, "Kepler's Theory of Hypothesis and the 'Realist Dilemma.'" In: *Studies in History and Philosophy of Science* 3 (1972), 233–264.
——, "The Comet and the Cosmos: Kepler, Maestlin and the Copernican Hypothesis." In: *The Reception of Copernicus' Heliocentric Theory*, ed. Jerzy Dobrzycki (Dordrecht: D. Reidel, 1973), 7–30.
——, "The Astronomer's Role in the Sixteenth Century: A Preliminary Study." In: *History of Science* 18 (1980), 105–147.
——, "Nature, Art, and Psyche: Jung, Pauli, and the Kepler-Fludd Polemic." In: *Occult and Scientific Mentalities in the Renaissance*, ed. Brian Vickers (Cambridge: Cambridge University Press, 1984), 177–230.
——, "Kepler's Early Physico-Astrological Problematic." In: *Journal for the History of Astronomy* 32 (2001), 227–236.
——, *The Copernican Question: Prognostication, Skepticism, and Celestial Order*. Berkeley: University of California Press, 2011.
Wilson, Curtis, "Kepler's Derivation of the Elliptical Path." In: *Isis* 59 (1968), 5–25.
——, "Horrocks, Harmonies, and the Exactitude of Kepler's Third Law." In: *Studia copernicana* 16 (1978), 235–259.
Yeomans, Donald K., *Comets: A Chronological History of Observation, Science, Myth, and Folklore*. New York: Wiley, 1991.

INDEX OF PERSONS

INDEX OF PLACES

INDEX OF SUBJECTS